WEBS AND SCALES

WEBS AND SCALES

Physical and Ecological Processes in Marine Fish Recruitment

Michael M. Mullin

Books in Recruitment Fishery Oceanography

Washington Sea Grant Program
Distributed by the University of Washington Press
Seattle and London

Publication of this book is supported by grant NA89AA-D-SG022 from the National Oceanic and Atmospheric Administration to the Washington Sea Grant Program, projects E/FO-3 and A/PC-5. The U.S. Government is authorized to produce and distribute reprints for government purposes notwithstanding any copyright notation that may appear hereon.

Library of Congress Cataloging-in-Publication Data

Mullin, Michael M. (Michael Mahlon), 1937-
Webs and scales: physical and ecological processes in marine fish recruitment / Michael M. Mullin.
p. cm. — (Books in recruitment fishery oceanography)
Includes bibliographic references and index.
ISBN 0-295-97244-0. — ISBN 0-295-97245-9 (pbk.)
1. Marine ecology. 2. Marine fishes—Ecology. 3. Food chains (Ecology) 4. Marine plankton. 5. Marine fishes—Larvae—Ecology.
I. Title. II. Series.
QH541.5.S3M85 1993
597.092—dc20 93-10188
CIP

WEBS AND SCALES

Physical and Ecological Processes in Marine Fish Recruitment

FOREWORD

The Washington Sea Grant monograph series, Books in Recruitment Fishery Oceanography, was established in 1988 with publication of Michael Sinclair's *Marine Populations: An Essay on Population Regulation and Speciation.* Alec MacCall's *Dynamic Geography of Marine Populations* followed in 1990 and William Pearcy's book, *Ocean Ecology of North Pacific Salmonids,* in 1992. The series, of which this is the fourth volume, is intended to communicate current thinking and findings in the field and to accelerate the synthesis of ideas.

Recruitment fishery oceanography is concerned with the effects of environmental variability on recruitment in populations of marine organisms, especially those of commercial importance. Such studies deal with the factors that determine the continuing productivity of living resources under environmental and fishing stress. They revolve around the intriguing scientific question of how population size is controlled in marine organisms.

The interactions of the control processes are complex but are slowly being disentangled. A short-term goal of this series is to improve the predictions of the level of catch that fish populations can sustain, so that fisheries can be managed beneficially. In the longer run, these studies also will lead to a better understanding of the consequences that natural or man-induced changes in climate have for marine organisms and their human predators.

Warren S. Wooster
School of Marine Affairs
University of Washington

Books in Recruitment Fishery Oceanography

PREFACE

Variability in the abundances of natural populations over time and space and the processes that cause this variability, or set limits upon it, are fundamental ecological issues. An assemblage of populations (and, implicitly, the ecological interactions between them) constitutes a community, whose central properties are the distribution of abundances among the constituent populations and the degree to which this distribution is maintained at a stable equilibrium. The nature and number of equilibrium states for a natural community are related directly to the numerical stability of its populations.

My intention in writing this small book, which is based on lectures given at the University of Washington in February 1989, and on several years of teaching with my colleague, John McGowan, is to emphasize, through examples and personal opinions, the confluence of two well-established lines of research aimed at understanding variability in marine, pelagic populations and communities. "Webs" refers to food webs in the plankton: the linkages between groups of organisms based on what eats what, with emphasis on determining rates of accumulation and transfer of energy and organic carbon. "Scales" refers both to ranges of space and time and to an emphasis on the ecology of larval fish (most of which are planktivorous) and their recruitment to the population of adults. (I generally use "recruitment" to mean reproduction and growth of the young stages of organisms with complex life histories into the general ecological rolès and habitats of adults, often with a metamorphosis, rather than in the fisheries sense, as defined by the size at which particular commercial gear captures them.) However, the issues to be addressed have equally relevant parallels in understanding variable recruitment into benthic invertebrate populations from meroplanktonic larvae, into submarine forests of large kelps arising from spores, and indeed into holoplanktonic populations.

Natural variability in populations, and its relation to climatic variability, has long been a source of stimulation and frustration for ecologists and a special, practical problem central to fisheries research. The possibility of accelerated climatic change in the next century, caused by human activities, has become a matter of public debate and legislative concern, and large research programs are being defined (or merely relabeled) in several countries. I have therefore tried to choose examples that not only present interesting scientific results and contribute to general understanding but also span the range of investigations that is being considered for these large programs. In doing so, I have also tried to identify some technical and conceptual impediments that must be overcome by special effort.

I have not attempted to describe the evidence that suggests that such climatic change will occur, nor the models and predictions of specific alterations in global circulation. I have, however, used the examples I chose as a basis for speculation concerning the implications of climatic change for the food webs leading to larval and juvenile fish.

As is already evident, I have written for readers who are familiar with the basic terminology of ecology, biological and fisheries oceanography, and physical oceanography. I have included liberal references to research as originally published, both to acknowledge some of those who have done work I find stimulating and to allow the reader to assess my interpretations independently. For pedagogic reasons, I have also mentioned shortcomings in some of these studies, but I have chosen the examples for their positive contributions and originality or clarity. I fully recognize that practical limitations, especially those posed by the ocean itself, impede the completeness, scope, and definitiveness that researchers strive to achieve. For similar reasons, I have shown actual data by using figures as originally published where possible, rather than clarifying the concept but suppressing the mensural variability through sketches.

To the extent that a complete picture of webs and scales emerges, it will be a montage constructed from these examples and my opinions, rather than a seamless whole or unified theory derived from fundamental principles or deduced from complete and unambiguous evidence. I hope that the montage will nevertheless provide new insights and perspectives and will clarify the terms of the ongoing debate as to the most effective ways to understand biological variability in the ocean.

Encouragement, guidance, and assistance from Warren Wooster and the University of Washington Sea Grant Program were essential to both the origin and the completion of this book. Three reviewers provided criticism that was both constructive and instructive, and Alma Johnson improved the text in many ways.

Although the phrase "another small block in the cathedral of science" is used rather cynically by researchers, it expresses the undeniable truth that science is a communal, cumulative enterprise, accomplished through (or occasionally in spite of) interactions of real people, with their real foibles and strengths. I acknowledge the contributions, some of which I do not even recognize consciously, of teachers, shipmates, lab mates, and my helpmate.

ACKNOWLEDGMENTS

Many colleagues and journal publishers have granted me permission to borrow the graphic products of previous research and thought; their names are cited in place. I have made much use of uncopyrighted materials from *Journal of Marine Research,* the reports of the California Cooperative Fisheries Investigations, and *Fishery Bulletin,* which is published by the National Marine Fisheries Service.

I gratefully acknowledge the permission of the following publishers to reprint materials (sometimes slightly modified) from copyrighted publications:

Academic Press, Inc. (Figures 2.1, Table 2.2)
Allen Press, Inc. (Figures 2.2–2.5, 3.1)
American Association for the Advancement of Science (Figure 4.25)
Bulletin of Marine Science (Figure 2.5)
Cambridge University Press (Figures 3.6, 3.7)
Danish Institute for Fisheries and Marine Research (Figures 3.9–3.12)
Department of Fisheries and Oceans for Canada (Figures 3.13, 4.26, 4.27, 4.30, 5.2–5.4)
Helgoländer Meeresuntersuchungen (Figure 4.1)
International Council for the Exploration of the Sea (Figure 4.18)
Inter-Research Publishers (Figures 2.6, 3.8, 3.12, 4.29)
Dr. W. Junk, Publishers (Figures 4.12–4.14)
Kluwer Academic Publishers (Figures 2.8–2.11, 3.21)
Macmillan Magazines Ltd. (Figures 2.7, 2.14)
Oxford University Press (Figures 2.7, 4.20–4.23)
Pergamon Press (Figures 3.16–3.20)
Plenum Press (Figures 3.4, 3.5)
Springer-Verlag Publishers (Figures 2.12, 2.13, 3.2, 3.3, 4.2, 5.1)
Westview Press (Figures 4.4, 4.6, 4.8)

1
SCALES OF PROCESSES AND INVESTIGATIONS

In fishery oceanography or fishery recruitment ecology, as in biological oceanography generally, the choice of a scale of investigation is usually based on an investigator's strong intuition as to what knowledge is most needed to advance the science at a given time or to rectify inadequacies in an existing paradigm. For example, if we cannot explain variation in recruitment solely by variation in spawning stock and in annual mean temperature, do we need to understand variability in feeding or predation at the individual level to make progress on the larger problem?

Related to this is the investigator's personal "scientific style," motivation, or source of employment, which may require a choice: either trying to develop a societally useful prediction in a short time or striving for insight into causation at a fundamental, mechanistic level, however long it takes. If large-scale, ongoing prediction is the goal, the kinds of data that can be used are often restricted to those that will continue to be readily available or are paid for by other elements of society (e.g., commercial catches), since a prediction based on a unique data set may not have ongoing utility.

Further, there is the issue of technical feasibility. New methods, often depending on technological advances in other disciplines, facilitate studies on new scales (e.g., the ability to rear larval fish and study individual behavior, or to determine daily growth rate from otolith rings or RNA/DNA ratio). There is also a fourth consideration, and it is sometimes an ignoble one: At what scale must one work to ensure finishing a Ph.D. thesis, justifying the next grant proposal, or earning tenure?

Finally, when an investigator has determined to do field work on a particular scale in the ocean, there is the mundane but essential problem of sampling. Since one usually extrapolates or interpolates from the scale on which measurements are made (e.g., samples taken on one day each month) to a different scale of interest (e.g., the seasonal cycle), a process on some third scale can introduce bias, contaminating the data record. For example, a small, anomalous patch that happens to be present on one sampling day may be interpreted as indicating a seasonal phenomenon. Also, the investigator must be careful not to alias the data by periodic sampling; the intersections between one scale of predominant environmental variability and another scale of sampling can give the entirely erroneous impression of a different scale of variability. This means that even though a particular scale is of interest, an investigator must sample so as to detect and correct for phenomena on other scales.

Both intuition into possible causes of variation and the need for predictions have led to investigations on large scale, in which some measure of recruitment (often, actually, catch of the youngest fish retained by commercial gear) is correlated with one or several environmental parameters, usually with some constant lag in time (that is, the success of recruitment is thought to be related to environmental events earlier in the life cycle). Unfortunately, the parameters are sometimes chosen more because data are available than because those parameters have the most likely causal potential. Even when care is taken in choosing parameters plausibly influencing recruitment, these are often themselves so intercorrelated on large scale that it is impossible, on statistical grounds, to identify a unique control of recruitment (i.e., a variation in a time series of recruitment data that is largely explained by only one parameter).

A more ecologically interesting question is whether such a correlation in data sets taken over one period of time or in one region, even if above reproach statistically, will be equally satisfactory at future times or in other regions. The relations could well be altered by genetic adaptations of the populations to their many interactions in any ecosystem (Rothschild 1986).

The other extreme is to synthesize an understanding of populations and communities from studies of individual behavioral or physiological ecology of "representative" organisms at small scale. This approach has obvious appeal for those who favor controlled experimentation (which is feasible on this scale), or who believe that the "scientific method" requires such control in order to test falsifiable hypotheses, or who are intellectually unsatisfied by the frequent ambiguity in proving causation from correlation. Further, it is sometimes argued that only penetration to the most fundamental, smallest scale—reductionism—will permit extrapolation to a prediction for environmental situations beyond those yet encountered, since the predictive capacity of correlations is often limited, at best, to interpolation within the range of previously experienced situations.

Investigators often focus on small scales for convenience, saving the larger scales for the discussion sections of their manuscripts, and sometimes rationalize this choice by comparing it to the physicist's search for ultimate particles or forces. (Fortunately, engineering solutions to real problems have not had to wait for the physicists' work to be completed.) But there is more than convenience or physics envy here. Ecosystems are complex and hierarchical in structure, and what is known of the general behavior of such systems also indicates the need to study small scales. An ecologist interested in the state of a large-scale system can often ignore or average out variations on smaller scales as long as the large-scale system is stable.

When the system is perturbed, however (as in the case of anthropogenic change in climate), the rapid dynamics of the small-scale components can become critical in determining how the larger system will respond. This is especially true as the system approaches a "bifurcation point" of major, relatively rapid, large-scale change (O'Neill 1988), where resilience breaks down. Thus, prediction of the large-scale change may require understanding of critical, small-scale variability. The trick is then to determine which of the multitudinous small-scale processes is critical, and how to concatenate the deductions concerning different scales.

It can be argued that small-scale processes may also have the opposite effect; that is, they may stabilize larger systems. For example, Steele and Henderson (1981) were stimulated by observations that pelagic communities in enclosures, even large ones, tended to vary in a quasi-deterministic fashion, while the community in the surrounding water was both more stable and more probabilistic. They considered that the variety of stochastic (or even cyclically varying) small-scale processes in the unenclosed water column buffer the system, preventing extreme changes in community composition. This theory of stability of planktonic communities has been termed "contemporaneous disequilibrium" (Richerson et al. 1970).

Thus, arguments can be made that both stability and change are governed by processes on scales in space and time that are small relative to the populations of interest, though not necessarily on the scale of interactions between individuals. In spite of this justification, the general weakness of many small-scale, carefully controlled studies is that while they often give considerable insight into what *can* or *may* happen on larger scales, they are usually incomplete in explaining what *does* or *will* happen in the full complexity of a natural ecosystem because they are inadequately linked to the larger scale. Obviously, studies of behavior and physiology have validity and significance within their own sphere. However, the issues in fisheries usually concern local populations or larger units that inhabit areas ranging from large embayments to major current systems. It is the extension of small-scale studies to larger scale—the ability to predict the demographic end product of behavior or physiology—that is at issue.

I have, of course, just described the extremes of a range of approaches. Many excellent studies are hybrids, and in the best of them, conclusions drawn from one scale of study are reinforced by supporting evidence from quite another scale. Also, it is by no means the case that only large-scale analyses are based on correlations between parameters the investigator does not or cannot control; processes on quite small scales are profitably studied in this way. Analytical or simulation models that are independent of actual data can be applied to processes on many scales, provided the physical processes dominating each scale are correctly represented.

Argument over the appropriate, or most productive, scales of study is not new in ecology, and it still continues. As recent examples, Hassell and May (1985) show through mathematical models how individual foraging behavior in an environment of patchy resources can interact with the degree and nature of patchiness to affect the stability of the user+resource system. Schoener (1986), using examples of successful predictions of niche separations and other community properties from studies of individuals (though involving only a few species), argues that community and population ecology are, in principle, reducible to individual ecology. However, he does *not* argue that such reductionism is necessarily the most efficient way to achieve a useful prediction, nor that variability is best studied in this way.

The debate also has a long history within aquatic ecology. An important feature of the pioneering studies of Gordon Riley (1946) was his specific contrast between a correlative approach to prediction of phytoplankton biomass based on large-scale field studies and one derived from the physiological ecology of those species that had been studied in the laboratory (very few, at that time). Good, recent discussions relevant to aquatic ecology are given in books by Harris (1986) and Carpenter (1988).

Multiple enclosures of large portions of natural aquatic communities—mesocosms containing tens to hundreds of cubic meters (e.g., Grice and Reeve 1982)—have been used in attempting to combine the advantages of experimentation with those of observation in a setting of near-natural complexity (though critics allege that the worst features of both approaches are achieved instead). Indeed, I believe that some early experiments in mesocosms stimulated more thought about the importance of small-scale hydrodynamics in maintaining the structure of planktonic communities than they taught us about the biotic interactions they were intended to illuminate. The model developed by Steele and Henderson (1981), reviewed above, is an example.

In fisheries, the debate may be exemplified by a recent book and its review. In the book, Rothschild (1986) states:

> A typical approach generally involves (a) selecting some time sequence of annual fish-stock abundance, (b) selecting some time sequence of abiotic or biotic environmental variable(s), and (c) analyzing the relationship between sequences . . . With such a procedure, it is not surprising that many correlations are found. But it is most difficult to distinguish between relations that are causal or predictive and relations that are simply correlative . . . The interaction of the physical environment with population variability cannot be understood by referring only to data on the environment and on recruitment . . . We should not be surprised . . . when correlations between recruitment and the environment deteriorate shortly after publication of the results . . . and we should be surprised when simple environmental correlations persist. When they do persist, it behooves us to explain how such simple relations override the compensatory mechanisms that must be inherent in the population!

The review (Collie 1988), though favorable, nevertheless says:

> Instead of taking a broad view, Rothschild, with a touch of physics envy . . . would have us study the encounter probabilities between fish larvae and their prey—the protons and electrons of ocean life. The rationale behind this reductionist philosophy is that we must understand how the micro-scale processes are integrated to produce the macro-scale changes in abundance that we observe.

In physical oceanography, the concept of a *spectrum* of scales, and the transfer of variation from one scale to another along this spectrum, seems well developed. That is, known hydrodynamic processes cause variance stimulated at one spatial/temporal scale to spread through the spectrum to other scales. There are peaks in the spectrum at various scales for fundamental reasons (e.g., the input of internal wave energy into the turbulence spectrum in a particular wavelength band). The nature of the hydrodynamics changes (in theory, if not in observational data) along the spectrum from the quasi-geostrophic, rotational regime at long wavelengths (Rossby number < 1.0, where Ro = velocity/[Coriolis frequency x length]) through the buoyancy and inertial frequencies at intermediate wavelengths to the viscous regime at shortest wavelengths (low Reynolds numbers, where Re = [length x velocity]/kinematic viscosity). Generally, however, we think of energy or variability associated with turbulence as cascading down from large to small scales; larger, more organized motions are generated by large-scale atmospheric and/or gravitational forcing and degenerate into smaller scales.

Biological oceanographers and fisheries scientists recognize that time and space scales should be measured relevant to particular organisms (e.g., generation time, daily ambit; see Steele 1978). Since the pelagic organisms humans care most about are rather long lived and mobile, the societally significant scales of variation are fairly large. Further, the implication for full-blown studies of a food web extending from bacteria and phytoplankton to fish is that many scales must be studied, since each kind of organism integrates its environment over a different scale. I think, however, that this has seldom been realized and is probably unreasonable to expect. We have few ecosystem studies that reveal the temporal variation per generation, or spatial variation within the daily ambit of an individual, of both phytoplankton and fish.

Since the dynamics of populations depend on the summation of outcomes of small-scale, individualistic interactions, it has been argued (as noted above) that knowledge of variation on the scale of an individual is therefore essential to understand variation in populations, i.e., that only through such knowledge can large-scale variation truly be understood. This reasoning is pervasive even though we have good evidence that the ocean tends to be biologically as well as physically "red" (i.e., dominated by variation on large scale at long wavelengths) rather than "white" (equally variable on all scales). The greatest variability occurs over very large areas and lasts a long time (Haury et al. 1978). Indeed, Steele (1985) has argued on theoretical grounds that the "redness" of physical variability in the ocean means that marine populations are likely to differ from terrestrial ones in their responses to climatic variation; the marine populations are more likely to show long-term ("red") variability because of the relation between typical generation times and characteristic scales of physical variability.

Theories relating the several scales of physical/chemical variation to biotic interactions within the food web (focused, if you like or if society so dictates, on particular species) still need to be improved. Technological (or financial) constraints also cause a mismatch between physical and biological oceanographic studies. Physical oceanographers now routinely use instruments that can measure significant properties such as temperature or current velocity not only very frequently and very close together in space but also for a long time over a large area; a spectrum of scales can readily be included in one study. Instruments that are self-contained, or are rapidly deployed and recovered (or even air-dropped), permit the ambiguity between spatial and temporal variation to be partially resolved. Satellite-borne sensors represent the ultimate tool for large-scale resolution, at least for surface properties. Having such a spectrum not only facilitates analysis of variability but also greatly reduces the likelihood of bias and alias in interpretation, mentioned earlier.

We have fewer such instruments in biological oceanography. Many of the devices we do have are unique to a particular program or investigator (see, e.g., Dickey 1988). Most do not measure that most significant of properties, species identity, much less physiological state, age structure, or genetic subpopulation.

The purpose of such instruments should be not only to acquire internally consistent ecological data on many scales, so that the magnitudes of variation on different scales can be compared, but also to permit such data to be analyzed in concert with physical/chemical data on the same spectrum of scales. For such interdisciplinary work to be most successful, interfacing of scales is a necessary but not fully sufficient condition. One also needs to try specifically to interface those scales in each discipline where predictive capability or understanding is greatest, given the state of knowledge in that discipline (O'Neill 1988). This, clearly, is not simple to achieve.

The intuitively reasonable converse of this proposition is that increased understanding on a particular scale in one discipline, such as fine-scale physics, should encourage the development of tools to gain equivalent understanding on that scale in allied or dependent disciplines, such as pelagic ecology. This is a natural and desirable evolution, even though it does not necessarily follow that the new understanding on one ecological scale will shed light on the ecological processes on other scales.

The issue of scales, and how to study them in different disciplines, is now brought to the fore by the evidence that anthropogenic, large-scale changes in atmospheric and oceanic climate are likely to occur in the near future. These changes apparently will be much faster, if not more drastic in magnitude, than those that natural ecosystems have experienced in geological time. Models are being developed to predict global winds and currents from the anticipated changes. We must consider the possibility that local- and smaller-scale distributions of properties and motions will also be affected, both through altered local wind patterns and through changes in the heat budgets and the distributions of density in the ocean, though it may be impossible to calculate them in detail (see, e.g., Longhurst 1984 and DeAngelis and Cushman 1990 concerning implication for fisheries). The conceptual problem will be to understand the nature of transition states lasting several decades, not just to deduce a new, stable distribution of properties in an equilibrium between ocean and atmosphere.

Because time, trained intelligence, and society's resources are scarce, biological and fisheries oceanographers must be judicious in choosing scales of investigations in order to understand the resulting variability in pelagic ecosystems. We must decide whether the ecological effects can be predicted best from correlations based on long time series of coarse-grained field data (i.e., data taken on scales large relative to the daily ambit or life span of individuals). We have, actually, rather few such sets of data for anything other than commercial catches and coastal climate, and one can argue that we need many more, for more elements of the food web. In this perspective, study of small-scale processes would be, in effect, irrelevant to predicting large-scale variation, or at best an inefficient way to proceed.

As the alternative approach, we must (a) deduce the implications of large-scale climatic change ("global" or "regional" models) for the physics (and chemistry) at the much smaller scale of individual organisms (micrometers and milliseconds for bacteria, kilometers and hours for fish)—change both in the mean conditions and in their variability; (b) analyze through controlled experiments the changes in physiology and behavior that these will cause ("process submodels"); (c) determine quantitatively all the critical interactions between species under these altered conditions; and then (d) reason back "up the spectrum" to variations in populations and communities by some weighted summation process and some food web paradigm. One problem with such linked models is that each is likely to be an incomplete or imperfect description of nature, even at its own scale, and we might well question the reliability of a prediction arising from concatenated uncertainties.

This dichotomy of approach is not so stark as I have painted it, nor are the choices so mutually exclusive. However, there is no question that our intellectual, technological, and financial resources for environmental research are limited and therefore must be deployed efficiently.

2
DISTRIBUTION AND PRODUCTION IN THE PLANKTONIC FOOD WEB: LARGEST AND SMALLEST SCALES

My purpose in this and the following chapter is to review, by means of examples, some of the distributions in space and time of components of the plankton and their rates of growth, particularly those distributions that can be related to specific scales of physical/chemical processes and that are likely to affect the food webs of larval and juvenile fish. I shall emphasize examples from the open ocean, though many of the strongest correlations between distributions and physical parameters are found in estuaries. As indicated in Chapter 1, the ultimate (but as yet unattained) goal should be to understand how the different scales affect each other, so that impacts of change in the physics and chemistry of the ocean on a large scale can be predicted on several biological scales. Ideally, these would also be the scales of greatest societal concern.

I shall set the stage by a brief description of "conventional" understanding of the large-scale, average distributions of planktonic biomass and production, as a contrast to what follows. I most definitely do not denigrate this understanding, nor do I imply that physical and chemical processes are less well integrated into it than into the studies I have chosen to review in greater detail. However, I shall explore more fully both the implications of some scales of physical and chemical processes that have been suppressed in the conventional descriptions (and, often, by conventional methods of measurement) and the question of whether these scales will be affected by changes in the ocean on larger scales. This summary is background to a consideration of whether variability on the scale of populations and communities is likely to be predictable from a summation of the effects of small-scale, altered processes. Unfortunately, such consideration does not provide a definitive answer.

I shall also review some mathematical models that simulate the effects of physical and chemical properties on biomass and production of plankton. The particular models are examples of a larger body of work; such models are usually unrealistic in their simplicity, but they permit the investigator to explore the consequences of a particular process in a way that even controlled experimentation does not. The best such models lead to conclusions that are both counterintuitive and testable with field data.

Ocean Basin and Regional Scales

The optimal large-scale conditions for rapid growth of phytoplankton in the open ocean occur in middle and low latitudes, where daylength exceeds eight hours all year, in those regions where the surface layers are only weakly stratified vertically. Regions in the higher latitudes can also support rapid growth, but only for the sunlit part of the year. These generalizations reflect the need of phytoplankton for adequate light and a supply of nutrients from waters below the euphotic zone (or, in some circumstances, from horizontal advection). In principle, grazing by zooplankton could keep phytoplanktonic biomass small by removing it as fast as photosynthesis produced it, but in fact the regions of rapid growth (on this large scale) also tend to have greater biomasses of phytoplankton than do less eutrophic regions, so the primary production (biomass-specific productivity times biomass) can be quite high (Table 2.1).

Table 2.1. Geographical variation in standing crop and primary production by phytoplankton

Type of area	Euphotic depth (m)	Chlorophyll (mg m^{-3})	Chlorophyll (mg m^{-2})	Phytoplankton biomass (g C m^{-2})	Gross production [mg C (m^2 day)$^{-1}$]
Strong upwelling[a]	< 30	10–25	100–400	5–20	500–4,000
Weak upwelling, mixing, benthic regeneration[b]	< 80	1–10	20–100	1–5	200–500
Weak mixing[c]	< 100	0.1–1	10–20	0.5–1	50–200
Minimal mixing[d]	< 150	0.05–0.1	5–10	0.25–0.5	10–50

[a]Benguela, Chile–Peru, other eastern boundary currents, some bays (runoff).
[b]Equatorial countercurrents, Gulf Stream, Oyashio Current, most shelves.
[c]Equatorial currents, tropical oceans.
[d]Sargasso Sea and other subtropical central gyres, Mediterranean.

Nitrate (NO_3^-) can be taken to represent all nutrients regenerated from sinking particulate matter by decay and (for N) subsequent oxidation. As a first approximation, it is supplied to the euphotic zone by turbulent diffusion or advection from deeper water. This provides the N for what has come to be called "new" production (which technically could include N in rainfall, etc.). Ammo-

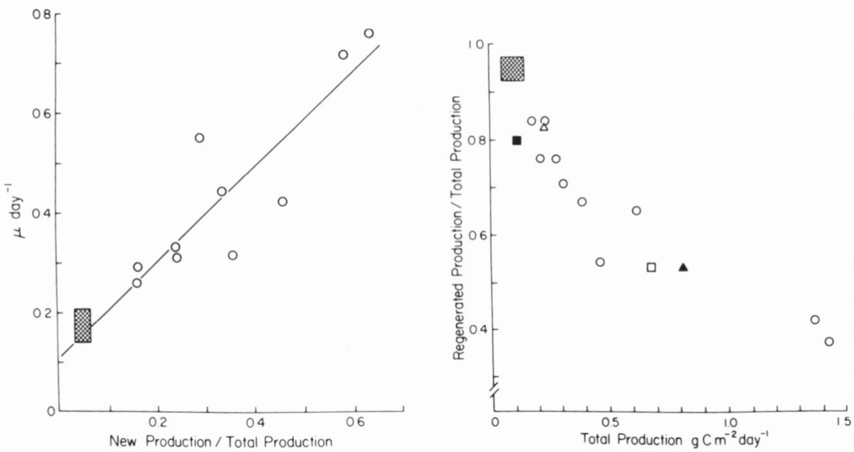

Figure 2.1 *Relations between growth of phytoplankton and source of nutrients.* Left: *biomass-specific growth rates (per day) of natural assemblages of phytoplankton vs. "f" ratio of new to total primary production.* Right: *regenerated (NH_4^+-based) as a fraction of total primary production vs. total primary production as carbon.* Open circles *are individual stations in the Southern California Bight,* shaded rectangle *is several stations in the North Pacific Central Gyre,* solid square *is Caribbean data,* open square *is offshore Monterey Bay,* solid triangle *is eastern tropical Pacific, and* open triangle *is eastern Mediterranean. From Eppley 1981.*

nium (NH_4^+) is regenerated within the euphotic zone as a result of metabolism and excretion by zooplankton and bacteria (Dugdale and Goering 1967). As a result, separate measurement of the uptake of these two forms of N illustrates the role of supply from below in fueling both high biomass-specific and total primary production of eutrophic waters. Figure 2.1 shows the relations between specific growth rate and the ratio of new to total primary production (sometimes called the "f" ratio) and the ratio of regenerated to total primary production as a function of total primary production as carbon.

Conversely, at large spatial/temporal scales, nitrogenous biomass that is removed from the euphotic zone, for example, when particles sink or fish are caught, must be replaced by "new" N if primary production and the food web are to continue to function (Eppley and Peterson 1979, Eppley et al. 1983, Horne et al. 1989). Interestingly, this concept is also an important one in considerations of change in global climate because of the so-called Redfield ratio, the intimate linkage between C and N in metabolism. The recycling of carbon within the euphotic zone, its sinking into deeper water, and the degree to which it remains there rather than returning to surface waters all affect how the ocean counteracts or augments anthropogenic changes in atmospheric CO_2 and the resultant "greenhouse" heating.

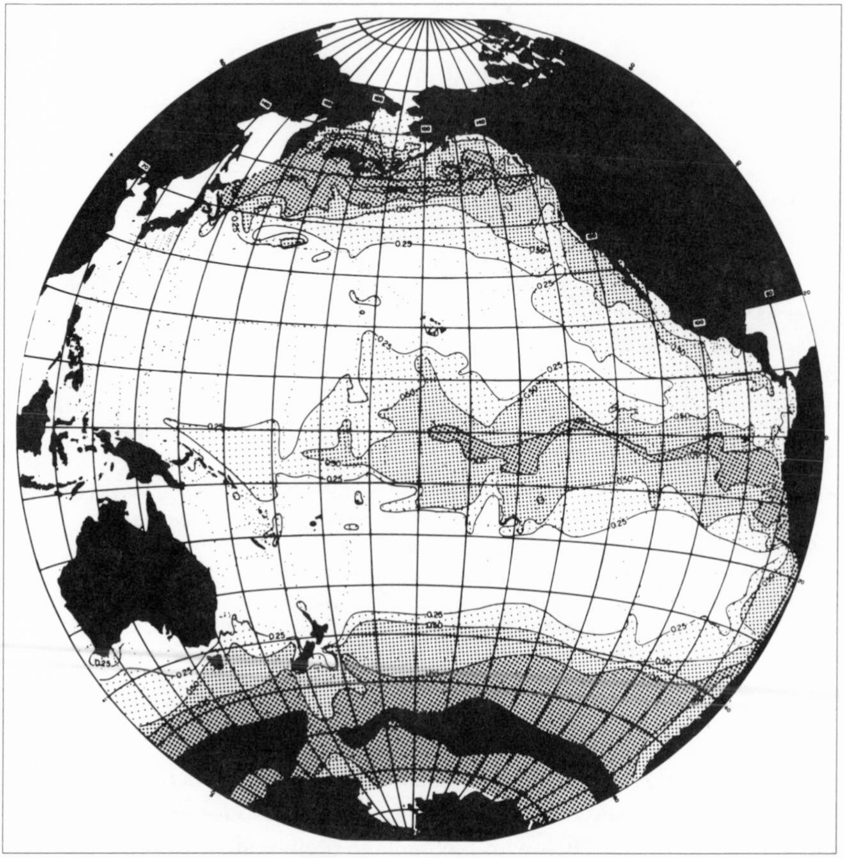

Figure 2.2 *Distribution of PO_4^{-3} (μM) at the surface in the Pacific Ocean. Darkest shading indicates values > 2 μM. From Reid 1962.*

As illustrated by the maps for the Pacific (Figures 2.2 and 2.3), zooplanktonic biomass in the upper few hundred meters also tends to be positively correlated with nutrients (phosphate; PO_4^{-3} in Reid's maps). These maps were prepared from data collected over several years, different areas being sampled in different years. The large-scale, general patterns are thus semipermanent, though the permanence of smaller features is unknown. In general, the areas that are eutrophic with respect to nutrients and plankton are also regions of the major pelagic fisheries (in terms of tonnage, if not value), though the distribution of fisheries is strongly biased by political, cultural, and economic constraints.

A major perturbation of the ocean climate such as the El Niño provides an illustration of the relation between nutrient supply and planktonic biomasses on large scale. Associated with the major Californian El Niño of 1958–59 were dramatic reductions of zooplanktonic biomass throughout the

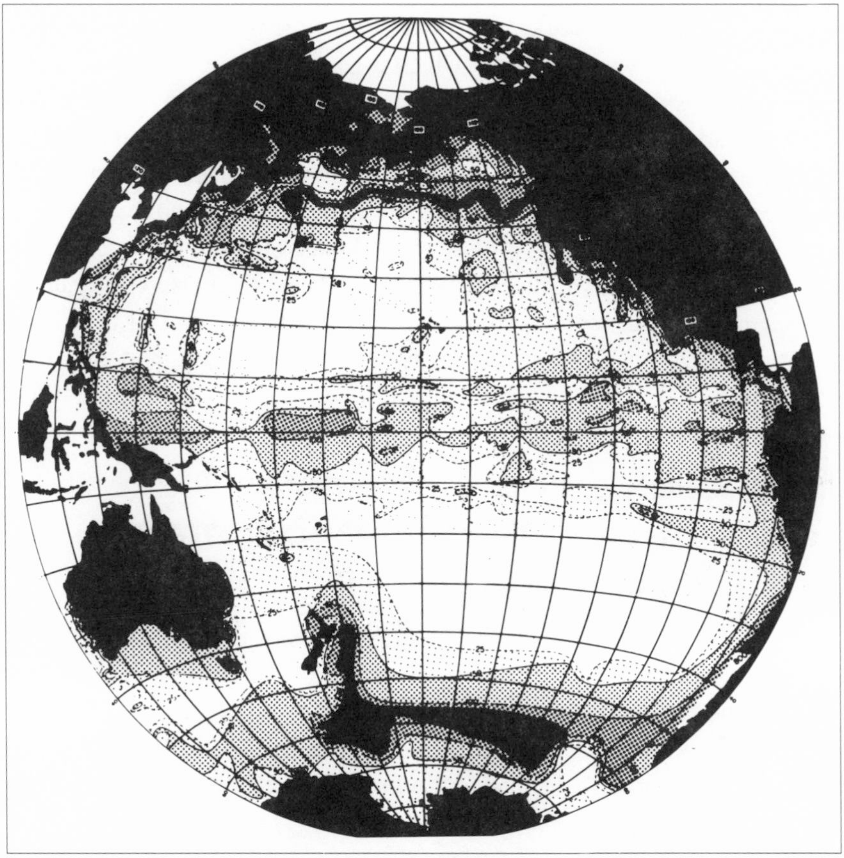

Figure 2.3 *Distribution of zooplanktonic biomass [ml $(10^3 m^3)^{-1}$] in the upper 150 m of the Pacific Ocean. Darkest shading indicates values approaching 1 ml m^{-3}. From Reid 1962.*

California Current (Reid 1962, Chelton et al. 1982). Properties of phytoplankton were not measured at that time, but more recent studies show dramatic reductions in phytoplanktonic biomass during the 1982–83 Californian El Niño. This was associated with a reduced incidence of cold, presumably upwelled water near Point Conception and marked a deepening of the nutricline, meaning that a greater portion of the euphotic zone was nutrient depleted (Fiedler 1984, McGowan 1985). Perhaps equally important for subsequent elements of the food web, less extensive sampling suggested that the size composition of the phytoplankton also was different in 1982–83, with very small cyanobacterial cells becoming a much larger component of the biomass and primary production (Putt and Prezelin 1985).

Exact causation is more difficult to establish, however. Evidence from fauna (i.e., "indicator" species) and a rise in sea level (indicative of altered geostrophic flow and thermal expansion) indicate a decrease in southward transport by the cool, nutrient-rich California Current and a penetration of more southerly or westerly water, which is generally more oligotrophic, into the region. It is not clear how much of the reduction in fertility is due to this change in advection, and how much to local reduction in vertical supply of nutrients to the euphotic zone (because of the thicker layer of warm, oligotrophic water at the surface). Roesler and Chelton (1987), who examined the timing of changes in transport and zooplanktonic biomass in the California Current, argued that the changes in biomass in the northern part of the current were largely advective, while those farther south (which lagged behind the changes in transport) were caused by responses to advected nutrients of the food web itself.

Microscales and Fine Scales Relevant to Individual Encounters and Microenvironments

Individual plankters interact with each other and, if they are phytoplankton cells, absorb nutrients on very small scales, from a few micrometers to a few millimeters and for no more than a few seconds. On the smallest spatial scale, where Reynolds numbers are small, flow is essentially laminar and viscous rather than turbulent. This significantly affects how nutrients reach the absorbing surface of cells and how zooplankters bring particles into their grasp. Correct description of actual mechanisms depends on correct scaling of the fluid dynamics. What is uncertain is the degree to which such correct descriptions will help us understand changes (or steady states) in populations and communities.

An individual phytoplankton cell is surrounded by a boundary layer of fluid through which nutrients (or bacteria) must diffuse to reach the outer membrane of the cell. As recognized by Munk and Riley (1952), a cell stationary in the water depletes nutrients in a microzone around it through absorption; sinking (or other motion relative to the water) minimizes this local depauperization. The size and shape of a particular cell affect both the severity of this depletion and the hydrodynamics of sinking. Even when uptake of nutrients is formulated as a saturable process (Gavis 1976), the issue is still a significant one.

Turbulence on the scale of tens of micrometers might then enhance uptake of nutrients (Munk and Riley 1952, Thomas and Gibson 1990a). However, there can be a negative aspect to such turbulence as well, at least for some kinds of phytoplankton. Thomas and Gibson (1990b) have shown that the cell-scale shearing stress that might be expected from turbulence at the sea surface under moderate winds can inhibit the growth of red-tide-forming dinoflagellates by causing loss of the flagellae with which these cells swim.

A phytoplanktonic cell not only takes up nutrients through the boundary layer but also excretes or leaks organic matter into it, organic matter that bacteria can utilize as a metabolic substrate for growth. Whether a bacterium can move close to a leaking cell by chemotaxis and remain in its vicinity, however, is a question of microscale physics (Mitchell et al. 1985). Jackson (1987) showed how the cell size and leakage rate of the phytoplankton and the behavior of bacteria determine whether there will be any effects on bacterial growth. The relevant point for the present discussion is that modeling of processes on a microscale can provide guidance in considering interactions between phytoplanktonic and bacterial populations.

Conventional sampling of phytoplankton involves catching on the order of 10^5–10^6 cells in a volume of 500 ml or so, drawn from a volume of several liters. Their bulk properties are then determined—uptake of $H^{14}CO_3^-$ for primary production, or content of chlorophyll or chemical composition, the latter confounded by nonphytoplanktonic particles present—as well as the concentration of nutrients in the same volume. For practical reasons, 3–10 such samples from various depths are assumed to represent the vertical conditions over a great horizontal and temporal range.

Studies of laboratory cultures have shown how the nutrient chemical concentration in cells and the growth rate increase with the external concentration of nutrients. Yet, phytoplankton sampled in oligotrophic waters with no measurable nutrients often has a chemical composition (and, by inference, a rate of growth) characteristic of plentiful nutrients. McCarthy and Goldman (1979) and Goldman (1984) have suggested that, since uptake is much faster than growth, rapid uptake by individual cells of nutrient micropatches (a few cubic micrometers) might supply sufficient nutrient for growth over a longer period (e.g., 24 hours). Such encounters would have to occur often enough to affect the bulk properties of the phytoplankton, but an instantaneous, large-volume sample cannot capture the process.

Other evidence of rapid phytoplankton growth in oligotrophic waters (Shulenberger and Reid 1981) lends credence to this suggestion. Experimental tests of uptake of NH_4^+ (an excretory product) support the idea, at least in part (e.g., Goldman et al. 1981, Horrigan and McCarthy 1982), and Lehman and Scavia (1982) presented evidence that excretion by individual zooplankters can create such usable micropatches. Jackson (1980) and Currie (1984a,

b) argued that, because of turbulent dissipation and nonlinear responses of uptake, excretory micropatches are unlikely to provide a significant number of cells with an adequate supply of nutrients. However, the idea is still attractive that microscale or small-scale processes are significant but cannot be detected by conventional sampling because of homogenization or statistical undersampling, or both.

In considering the larger issue of patchiness of nutrients, it is worth remembering that variability in nutrient concentrations may play a role in floral diversity as well as in the overall rate of growth of the phytoplanktonic assemblage. Species differ in the kinetics (rate and saturating concentration) of uptake of nutrients, so that temporal variation may permit several "kinetic types" to coexist, even when all are apparently limited by the same element (see, e.g., Grenny et al. 1973, Turpin and Harrison 1979). Floral composition may affect the pathways of primary production, whatever its magnitude, through the food web to juvenile fish and therefore represents, for the fisheries oceanographer, a second facet of the question of nutrient patchiness.

At slightly larger scales, organic aggregates and the physical/chemical microhabitats they create become significant because they are an important exception to the classic view of plankton as organisms suspended individually in a relatively homogeneous environment. They are potentially relevant to the foregoing discussion because they might provide a physical barrier retarding the dissipation of the metabolic byproducts of their microcommunities. By "organic aggregates," I refer to assemblages of heterogeneous organisms and other seston, loosely bound together in a detrital matrix of organic origin. Because the larger aggregates can be individually collected by divers, they have been studied the most. Such aggregates vary in size (several micrometers to several centimeters), shape (sheets, flakes, flocs, fibers, diaphanous webs), texture (mucilaginous, fibrous), and origin (cast exoskeletons, mucous feeding structures, feces, bacterial films, clumps of chain-forming diatoms). The common feature is the relatively long term proximity of the associated organisms and the chemical microhabitat created by their metabolism. Alldredge et al. (1990) showed that aggregates derived from larvacean houses are sufficiently robust to resist disruption by normal turbulent energy dissipation rates.

Phytoplankton blooms are traditionally thought to end by exhaustion of nutrients and grazing by an increasing stock of herbivorous zooplankton. Occasionally, however, residues of blooms are found as flocs of apparently uningested phytoplankton on the seabed, rather than as fecal pellets. Jackson (1990) has argued, from considerations of hydrodynamic events on the microscale, that the transition from a developing bloom of individual phytoplanktonic cells to a postbloom condition of sinking aggregates made up of adhering cells can be quite sudden, even without the increase in sinking rate of individual cells that is known to accompany depletion of nutrients. These diatom flocs, while still abundant in the water column, are the type of aggregate most likely to contribute more than a small fraction to the summation of metabolic processes there (Alldredge and Gotschalk 1990).

The microenvironments of aggregates can be ecologically important if the metabolic byproducts are themselves nutrients for other organisms. For example, Shanks and Trent (1979) and Alldredge and Gotschalk (1990) reported elevated concentrations of NH_4^+ in the immediate vicinity of aggregates, and Gotschalk and Alldredge (1989) demonstrated buildup of NH_4^+ around aggregates enclosed for several days. Primary productivity, however, is not necessarily enhanced (Alldredge and Cox 1982, Prezelin and Alldredge 1983), and the biomass of phytoplankton associated with aggregates (at least macroscopic ones) is usually insignificant relative to the overall biomass in the water column (e.g., Alldredge and Cox 1982, Beers et al. 1986, Alldredge and Gotschalk 1990), although the concentration of phytoplanktonic biomass on aggregates usually exceeds that in the water.

The physical surface of an aggregate, as well as its chemical nature, may make it a site of greatly elevated concentrations of bacteria and heterotrophic protozoans (e.g., Caron et al. 1982). Mitchell and Fuhrman (1989) invoked this possibility to explain fine-scale patchiness of bacteria sampled blindly. Large-scale distributions, such as onshore-offshore gradients, may be different for organisms associated with aggregates than for those free in the water (Caron et al. 1986). Indeed, there may be on microscales a pattern of succession in the microbial community living on aggregates (though not necessarily restricted to them), a pattern that results from processes on larger scale (e.g., plankton blooms) which enhance the formation of the aggregates (Davoll and Silver 1986). Net consumption of O_2 by aggregates, even in the light, suggests that heterotrophic processes exceed photosynthesis (Alldredge and Cohen 1987). However, on a larger scale the concentration of free-living bacteria so far exceeds that of bacteria associated with macroscopic aggregates that the free-living bacteria dominate heterotrophic processes in the water column, even though the rate of heterotrophy per bacterium may be significantly enhanced on the aggregates.

Aggregates significantly affect the concentration and predator–prey interactions of the organisms that occupy them. By aggregating living and dead particles of many sizes and types, they also affect the transport of particulate organic matter from the euphotic zone to greater depths. However, the influence of this group of phenomena on macrozooplankton and larval fish, either directly as food or indirectly, is uncertain. It would be interesting to determine, by analysis of gut contents with appropriate antibodies, the extent to which larval fish benefit from the microbial food web on aggregates by ingesting heterotrophic microflagellates that are too small to be captured individually.

Feeding by herbivorous copepods has been studied both because it is a source of mortality for phytoplankton and because it provides nutrition for a major food of small fish. The process was once viewed as a mechanical filtration, in which water is driven through a filter of setae or the filter is raked through the water, and cells caught on the filter are transferred to the mouth for ingestion. The mechanisms of capture of food particles are now better known, thanks to the studies of Strickler and others (e.g., Strickler 1982, 1984, Paffenhöfer et al. 1982, Price et al. 1983), and to the realization that, because of the fluid dynamics on this scale, such filtration is impractical in viscous media. Many particle-grazing copepods create some sort of flow to bring water and phytoplankton near their mouthparts, but Strickler's films reveal that at least some copepods sense the presence of an individual large cell before contact, and alter the motion of the mouth parts so as to draw the cell towards the mouth. Ancillary evidence suggests that copepods "smell" (i.e., sense chemically at a distance) the approaching cells; Andrews (1983) has modeled how a sphere of scent around a cell is distorted in the feeding current created by the copepod, and Paffenhöfer and Lewis (1990) determined by microcinematography the distances at which copepods react to cells.

Traditionally, experiments to determine the rates of grazing and ingestion by copepods have been conducted in closed containers that are either unstirred or stirred at some arbitrary rate to keep the particles of food in suspension. Rothschild and Osborn (1988) have correctly pointed out that these conditions may be quite artificial relative to natural turbulence, and they demonstrate mathematically that turbulence may increase the rate at which copepods contact cells by 20% or even 50%, relative to the rate in stagnant water. This may be important when the cells are too rare to saturate the mechanism of capture and ingestion. Simulation of the motions of swimming predators and prey in a turbulent environment also demonstrated that contacts increase, especially if the radius within which the predator can detect and capture the prey is fairly large (Yamazaki et al. 1991). Examination of the rate of ingestion by cod larvae of copepod nauplii at various concentrations under conditions of different wind-induced turbulence has shown up to a twofold increase in feeding caused by turbulence in this predator–prey situation (Sundby and Fossum 1990; see also Chapter 4).

Much attention has been paid to ingestion as a function of concentration of food and to the small-scale patchiness of food (see, e.g., Mullin and Brooks 1976). Small-scale turbulence can be quite patchy in time and space (e.g., Yamazaki and Osborn 1988), and Rothschild and Osborn's results suggest that one should consider this patchiness as well.

However, turbulence and patches of plankton are antithetical. If patches of anomalously dense concentrations are important places for feeding, as indicated by the model constructed by Davis et al. (1991), planktivores can be well nourished either when absence of turbulence permits small-scale patchiness or when turbulence is sufficient to enhance markedly the rate of encounter between an individual predator and its prey. Intermediate levels (or scales) of turbulence may be the least advantageous for the predators.

It is also possible that change in the large-scale physical circulation of the mixed layer is accompanied by changes in the distribution and intensity of small-scale turbulence, and therefore affects the grazing of copepods. However, I doubt that this change, even if extreme, would outweigh the concurrent changes experienced by the food web of the copepods, such as altered advection and altered supply of nutrients mixing into the euphotic zone from deep waters.

Small-scale Vertical Distributions

Some of the earliest investigations in pelagic ecology concerned the vertical distributions of plankton and how, for many species of zooplankton, these distributions change dielly, seasonally, and ontogenetically on scales of tens to hundreds of meters. While there are still valid questions concerning the proximate and ultimate (i.e., evolutionary) causes and consequences of such distributions, I am here concerned with vertical distributions on smaller scales, such as those within the euphotic zone.

Even in situations in which physical and chemical properties are uniformly distributed with depth in the upper few tens of meters, forming the so-called mixed layer above the thermocline, plankton may be patchy (i.e., nonrandom) in distribution. In some cases, the horizontal extent of such features is great enough to consider them layers, but information is generally insufficient to establish this because even data taken continuously in the vertical are generally from stations that are widely separated horizontally. Underway acoustic data frequently show quasi-continuous layers of micronekton, but they are usually far below the thermocline, at least by day, and are made up of large, mobile organisms. However, a sound-scattering layer extending across the North Pacific at depths shallower than 100 m day and night, dominated by the copepod *Calanus (Neocalanus) cristatus*, was reported by Barraclough et al. (1969).

The question of vertical patchiness of larval fish food is of concern because larvae reared in the laboratory often require concentrations of food which considerably exceed the average concentrations in surface waters. Success in early larval life might therefore depend on feeding in patches of anomalously high concentrations of food—either that, or the rearing experiments have been misleading. (By "anomalous patches," I mean that regions with high concentrations are more frequent than would result from a random distribution of abundances about the mean.) Whether such patches exist in the mixed layer, how long they persist, and whether they are associated with physical features that might increase the probability of larval occurrence, are therefore significant issues. I shall return in Chapter 4 to the importance of patches and layers of anomalously high concentrations of plankton for first-feeding larval anchovy; here I review a few other examples and their implications for larval fish generally.

Owen, in a recent study (1989), has observed patchy distributions of plankton within vertical distances of 2 m, which is essentially on the fine scale. The coefficients of variation (standard deviation/mean) on this scale for various types of plankton often exceeded those on the scale of 1–10 m, while nutrients and temperature were more variable on the larger scale (but still much less so than the plankton). Particular types were often patchier than were measures of total biomass such as chlorophyll or total particles.

On this scale, then, the composition of potential prey for larval fish was more variable than the biomass, but in fact the types thought to represent larval food were less patchy than the potential competitors or predators on the larvae. The physical stability of the water column may have influenced patchiness, since vertical profiles tended to be least patchy at wind speeds greater than 10 m sec^{-1}, in the "mixed" layer rather than in the thermocline, and in shallow areas rather than over deep water. How much of the observed patchiness might be due to the association of particular organisms with relatively rare organic aggregates is not known.

Patchiness was greater diurnally than nocturnally in Owen's study. Although the mechanism was not established, this finding is consistent with nocturnal cooling and overturn of surface waters; and since larval fish are daytime feeders, perhaps it is fortunate that this is the case. Certainly the implication of Owen's study is that the distributions of prey and predators on the scale that a larval fish might traverse in a few hours can be quite variable. It is unfortunate that larval fish were too scarce to be sampled on this scale as well. It is also unfortunate that the horizontal dimension characteristic of the patches was not established, since this affects their potential permanence and the likelihood that a larval fish, once finding a patch, could feed there for many hours.

A striking example of small-scale patches or layers, though based on only a single station and time, is reported by Bjornsen and Nielsen (1991). Within a vertical range of 2 m around a sharp pycnocline, a larval fish could have encountered very high concentrations of a toxic dinoflagellate, *Gymnodinium aureolum*, and few edible protozoans, or significantly more abundant oligotrichous and tintinnid ciliates above and below the layer of maximal *Gymnodinium*.

Rearing of copepods in the laboratory has led to a conundrum analogous to that arising from larval fish rearing: the mean concentrations of phytoplankton do not seem to be sufficient for survival and growth. The problem is complicated by the fact that we do not really know how to measure the food supply of a naupliar copepod in nature. Is it total chlorophyll, total particulate organic carbon, total living carbon (as estimated from, say, ATP), or some particular size category or biochemical subset of one of these?

Because gradients in phytoplanktonic biomass (i.e., change in biomass per unit distance) are much stronger vertically than horizontally (e.g., Mullin 1986), much attention has been paid to whether zooplankters occur and feed in layers (or vertical patchiness) of anomalously high concentrations of food. Mullin and Brooks (1972, 1976) reported examples from the Southern California Bight in which the abundance or biomass of the copepod *Calanus pacificus* was positively correlated with biomass of phytoplankton (though with great variability; see Figure 2.4), due primarily to vertical distributions on the scale of meters to a few decameters. Although the mean concentration of phytoplankton was insufficient for growth at many stations, distributions of "malnutrition and surfeit" during the two periods studied could be calculated from the specific vertical distributions and laboratory data on feeding and respiration. Cox et al. (1983) calculated similar distributions for two portions of the Bight, one richer than the other, from similar data taken in another year (Figure 2.5).

These relations result from sampling, at different times and locations, of the vertical distributions of a copepod species and its presumed food (or, at least, a property, chlorophyll, presumed to be correlated with food). The

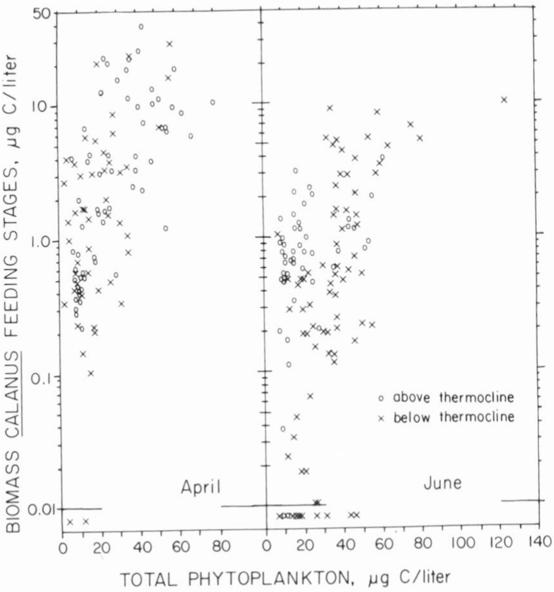

Figure 2.4 *Biomass of* Calanus pacificus *and phytoplankton at various stations and depths in the upper 50 m of the Southern California Bight on two occasions. See Figure 2.5 for metabolic implications of distributions. From Mullin and Brooks 1976.*

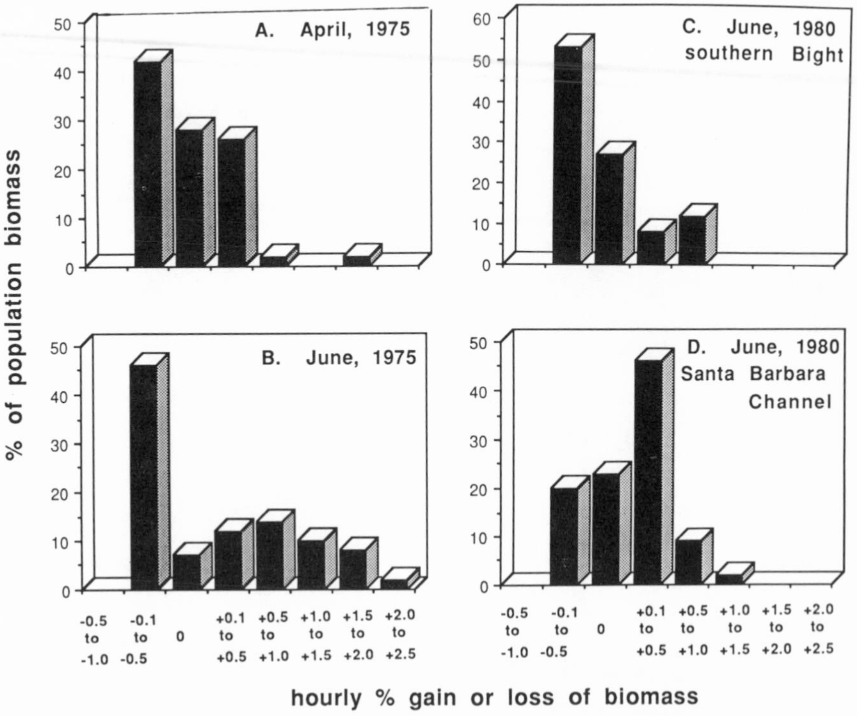

Figure 2.5 *Distributions of fractions of the biomass of the* Calanus pacificus *population in the Southern California Bight experiencing different percentages of hourly losses or gains of biomass at four times or locations, calculated from ingestion, assimilation, and respiration at several stations and depths. For correlations with* A *and* B, *see Figure 2.4. Modified from Mullin and Brooks 1976, and Cox et al. 1983.*

questions yet to be answered concern the roles of behavior of the copepods, the imbalances between reproduction and grazing on phytoplankton, and turbulence in restructuring the vertical distributions at one location. Mullin et al. (1985) studied vertical distributions before and after a local storm, and found surprisingly little disruption. (This observation will be discussed more fully in Chapter 4.)

While it is tempting to invoke the use by zooplankton of patches or layers of food as the "solution" to the apparent insufficiency of average concentrations to sustain growth, the range of natural situations, even within one geographical area (the Southern California Bight), is much more complex. First, the variability shown in Figure 2.4 means either that the relation between phytoplanktonic carbon and quality of food is quite variable (on this issue, see Mullin and Brooks 1976, with regard to cell size, and Napp et al. 1988a, concerning biochemical composition) or that the copepods are not very efficient at finding the best places to feed. Second, there is at least one example in which *Calanus* and other zooplankters seemed almost to avoid a dense layer of food (Fiedler 1982). Third, Napp et al. (1988b) found no relation between the vertical distribution of many species and that of phytoplanktonic biomass. Finally, Dagg and Wyman (1983) and Mullin et al. (1985) were unable to show that there was a relation between the vertical distribution of chlorophyll and the depths at which copepods fed most intensively (as reflected by plant pigments in their guts).

Vertical Motions and Their Effects on Primary Production

Because of the vertical attenuation of light in the sea (which as a first approximation is an exponential decay) and the reservoir of oxidized nutrients represented by deep water, the scale of a few meters to a few tens of meters below the surface is crucial for the physiology of phytoplankton. In stable situations, cells photoadapt; that is, cells living near the surface have less chlorophyll than those living near the base of the euphotic zone, and different kinetic responses to light. The nitrogenous nutrient used by phytoplankton near the surface is likely to be ammonium excreted by heterotrophs, while deeper-living cells often use nitrate (depending on the penetration of light relative to the depth of the nutricline, where nitrate begins to increase with depth). Finally, there may be distinct communities of species at different depths, particularly if the euphotic zone extends below the thermocline (e.g., Venrick 1988).

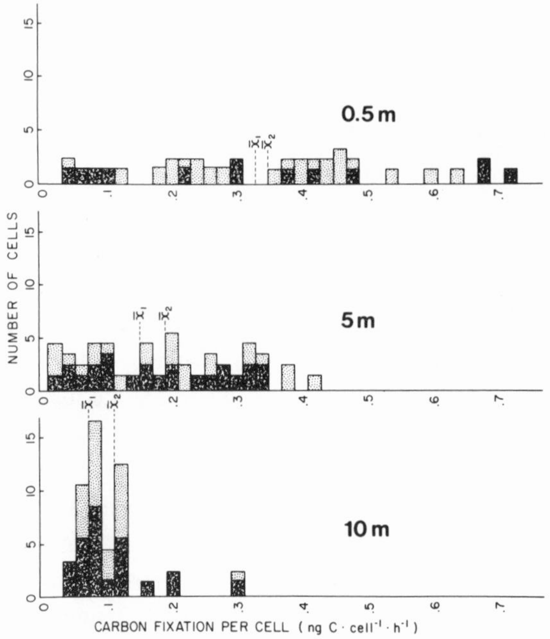

Figure 2.6 Primary productivity of individual cells of Ceratium tripos *at three depths. Shadings denote two water samples. Mean productivity per cell of each sample is given. From Boulding and Platt 1986.*

All this means that both conceptual and simulation models of the euphotic zone as a vertically stratified column are well developed (e.g., Jamart et al. 1977). In general, these models consider the phytoplankton at any depth to be physiologically homogeneous (i.e., to respond similarly, having had similar previous exposure to nutrients and light), and to move vertically only by sinking (or, in the special case of some dinoflagellates, by swimming vertically). Most measurements of primary production at sea are based on this conceptual model.

Nevertheless, there is good evidence that the individual cells at a given depth and time do differ in physiological properties. Micro-autoradiography of samples incubated with ^{14}C has shown that species within less than a liter of water differ in their photosynthetic rates (e.g., Maguire and Neill 1971, Knoechel and Kalff 1976a, b). Even within a species, photosynthetic uptake may differ considerably (Figure 2.6). The possible sources of this variability are the sizes and ages of the cells, their environmental histories (they may have been brought together from different depths by physical mixing), and their genetic makeups.

Denman and Gargett (1983), in reviewing the processes resulting in vertical motions in surface waters, evaluated each in terms of characteristic time and space scales, relative to the time course of adaptation of photosynthesis. Turbulent mixing and internal waves were the most important motions

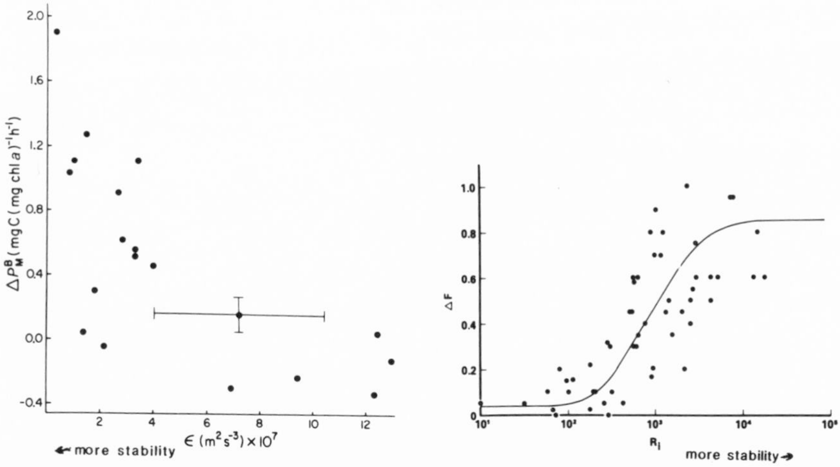

Figure 2.7 *From two studies, differences in photosynthetic responses between surface-living and deep-living phytoplankton, plotted against rate of mixing.* Left: *difference between maximal photosynthetic rates of surface and deep samples as a function of turbulent dissipation rate (higher values mean more mixing). From Lewis et al. 1984.* Right: *difference between relative inhibition by* DCMU *of photosynthesis of samples from deep water (where the effect tends to be greatest) and inhibition at the surface (higher values mean greater difference between depths) as a function of the Richardson number (lower values mean more mixing). From Harris 1984.*

in terms of their effect on primary production.

As noted above, differentiation in the kinetics of primary production can arise in a stable water column in cells growing in different microenvironments, from some combination of species differences and physiological adaptations. There ought, therefore, to be measurable effects of the relation between the speed of physiological differentiation and the rate of mixing. Figure 2.7 shows the results of two studies in which a measure of the difference in photosynthetic responses between surface- and deep-living phytoplankton is plotted against a measure of the rate of mixing. Although the relations look superficially different because different properties are plotted, in fact both graphs show that at high rates of mixing the responses are similar, and at low rates there are differences between the two populations—or, more correctly, between the communities, since the specific composition may have differed vertically, also.

It would be interesting to use micro-audiography or the technique of Boulding and Platt (1986) to learn the degree of physiological differentiation within single depths in these studies. One would expect that, as turbulent mixing increases to some intermediate range, within-depth heterogeneity increases and between-depth differentiation diminishes, as cells with different histories (and/or of different species) are increasingly brought together

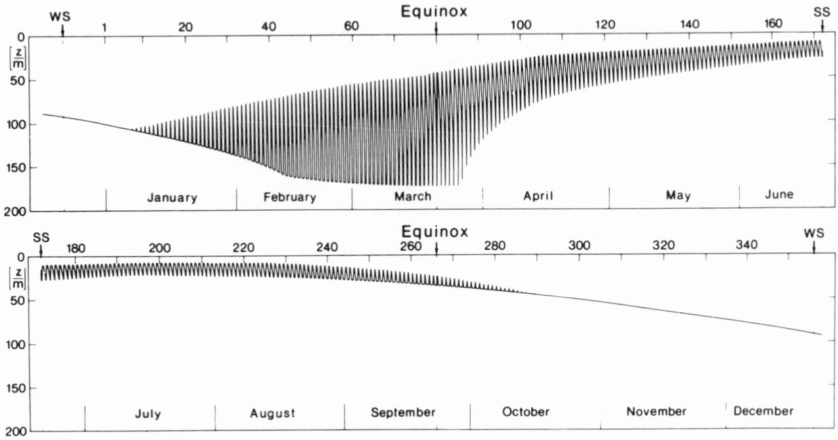

Figure 2.8 Idealized behavior of the midlatitude thermocline through seasons, affected by diurnal heating and nocturnal cooling. From Wolf and Woods 1988.

(Falkowski and Wirick 1981). Gallegos and Platt (1982) designed an interesting incubator to mimic this, but they were unable to demonstrate ecologically significant effects.

A special form of vertical motion which is seasonal in occurrence but diel in period results from the diurnal heating and nocturnal cooling at the surface, which creates a diel deepening and shoaling of the thermocline (Figure 2.8 is an idealized view). Wolf and Woods (1988) have used a Lagrangian ensemble model to simulate the temporal sequence of the bulk property, biomass of chlorophyll. The model explicitly retains information on individual cells with different histories that can co-occur at one depth and time due to random mixing above the thermocline. In principle, at least, such a model should be more realistic than those that treat all the cells in any layer at one time as physiologically homogeneous. In this context, the term "thermocline" may be misleading. What is important is the diel variation in the penetration depth of mixing, which may differ from the depth of the thermocline as a biologist would define it.

In the Wolf and Woods model, if a cell is left behind in the deep, stable layer in the morning as the thermocline shoals, the cell sinks and may or may not be overtaken and re-entrained into the mixing layer as the thermocline deepens in the evening (Figure 2.9). The photic history and nutrient uptake of each cell depend on its location in the water column, and a given cell (or its daughter cells) can vary between nutrient limitation and light limitation over time. Uptake by cells depletes the nutrient pool, and there is no regeneration in the model.

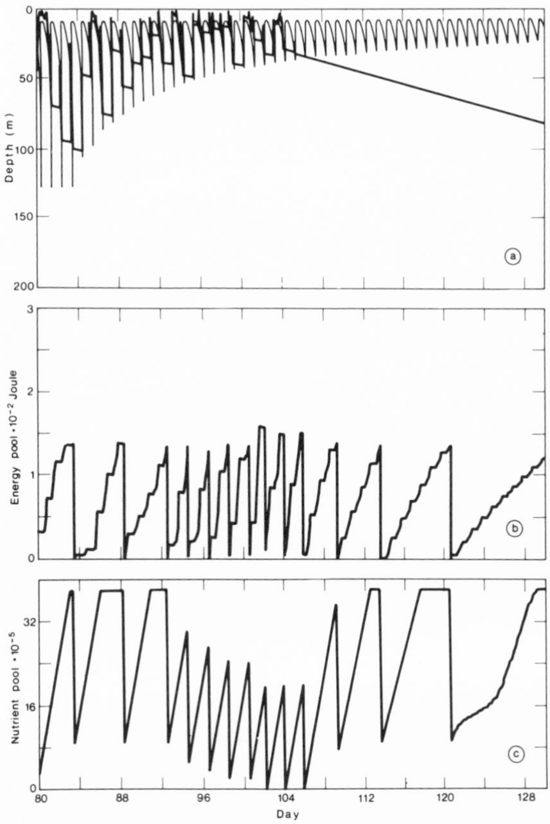

Figure 2.9 Individual cell, mid-March through mid-May: a, *trajectory through time and depth of an individual cell relative to the thermocline in Figure 2.8;* b, *energy content;* c, *nutrient content of this single cell, increase being due to photosynthesis (daytime only) or uptake and decrease being due to cell division. Note that rate of division is greatest when trajectory is confined to upper 50 m. From Wolf and Woods 1988.*

When the vertical distributions of nutrients and chlorophyll are determined as a function of time, a realistic formation and deepening of the nutricline and chlorophyll maximum layer results (Figure 2.10). In fact, this is not too different from the outcome of the simpler model of Jamart et al. (1977), in which variation in sinking rate as a function of nutrient uptake was a significant element; but the Lagrangian model might be more realistic in mechanism and, therefore, a better tool with which to examine the effects of perturbations. Even a single day of deep mixing, such as might be caused by a storm, can alter the vertical distributions significantly (Figure 2.11), and one can imagine a local storm thus creating horizontal patchiness in its wake. Less severe wind conditions, such that Langmuir cells become the dominant agent of mixing, should also be considered.

The first step in assessing the generality of this model is to obtain more data sets demonstrating the circumstances resulting in a diel variance in the thermocline idealized in Figure 2.8. An informative further test would be to conduct an analysis of variance in floral composition and in physiological

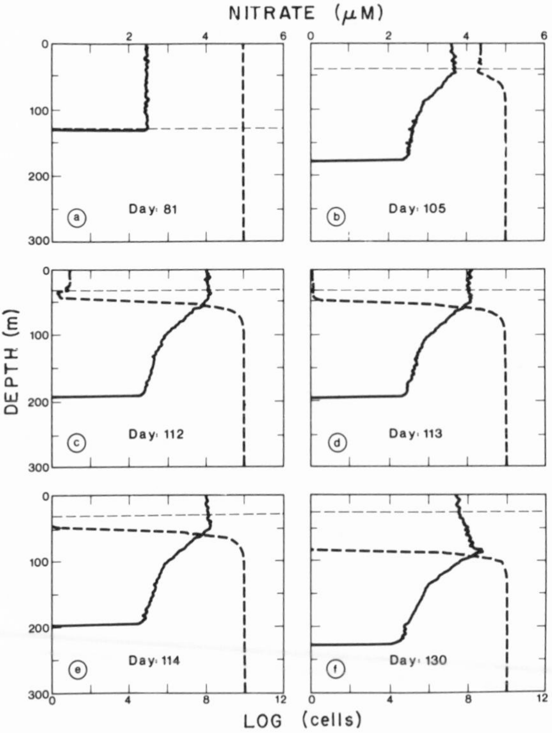

Figure 2.10 Vertical distributions of NO_3^- (dashed line) *and phytoplankton abundance* (solid line) *from day 81 to day 130 of a simulation. Note that the axis for phytoplankton is logarithmic, de-emphasizing the maximum. The horizontal dashed line indicates the deepest depth of the mixed layer in the past 24 hours. From Wolf and Woods 1988.*

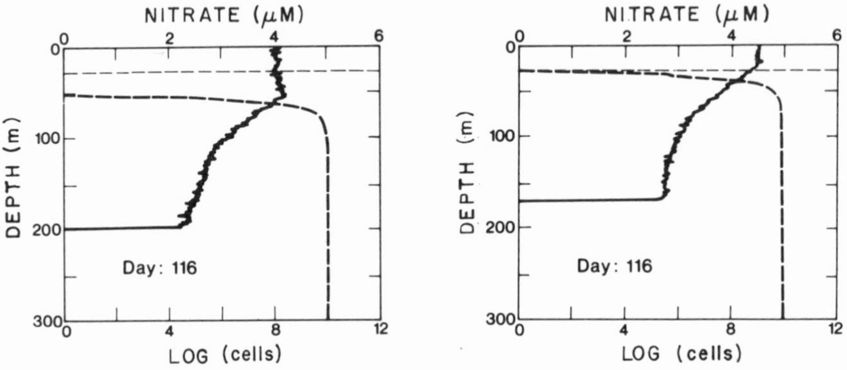

Figure 2.11 *Vertical distributions of* NO_3^- (dashed curved line) *and phytoplankton (log scale,* solid line): left, *without upwelling;* right, *effect of upwelling event. Compare with Figure 2.10. From Wolf and Woods 1988.*

states of phytoplankton as a function of season, using methods such as the example by Boulding and Platt (1986) to see whether the variance at a particular depth was indeed greatest, and the floral assemblage most vertically similar, during the season of diel oscillation of the thermocline.

Stratification of the water column reduces turbulent mixing but promotes sharp density interfaces on which internal waves can propagate. On continental shelves, such waves are pronounced at the 12.4-hour period of the semidiurnal tide, while in the open ocean, periods range from a few minutes (the local buoyancy period) to about 17 hours (the local inertial period), with crest-to-trough vertical displacements at the longer periods of up to 30 m (Denman and Gargett 1983).

For cells living near interfaces such as a sharp thermocline, these vertical motions expose them during the day to exponentially increasing or decreasing intensities of light. If the interface is in the lower half of the euphotic zone, the photosynthetic rate is approximately linearly related to light intensity, and so the effect is a net augumentation of photosynthesis over what the cells would experience at rest.

Table 2.2 (from LeFevre 1986) shows the effect on illumination of cells in the middle of a chlorophyll maximum layer 2 m thick riding on internal waves of 4 m amplitude at a mean depth of 25 m. The water above the chlorophyll maximum has an attenuation coefficient of 0.1 m^{-1}, and the attenuation in the maximum layer is 0.15–0.6 m^{-1}. The fraction of surface illumination received by these cells is compared with that received by cells in an on-deck incubator adjusted for the mean depth (24 m of surface water plus 1 m of the maximum layer). Note that, because of the exponential attenuation of light, this is somewhat less than the average of the intensities experienced by the cells on the internal wave. Also shown in the table are the intensities experienced by cells in a bottle suspended in the sea at 25 m on a taut line, where the cells actually experience greater illumination when the trough of the wave passes because this moves the turbid maximum layer below the bottle.

Table 2.2. Ambient light, as percentage of surface value, experienced by cells in the middle of a chlorophyll maximum layer 2 m thick centered at 25 m, compared with ideal on-deck and *in situ* incubations.

Chlorophyll maximum attenuation coefficient	State of internal wave	Illumination (percent) In nature	Deck incubator	Suspended *in situ*
0.15 m^{-1}	crest	11.6	7.8	7.4
	trough	5.2		8.2
0.4 m^{-1}	crest	9.1	6.1	4.5
	trough	4.1		8.2
0.6 m^{-1}	crest	7.4	5.0	3.0
	trough	3.3		8.2

Lande and Yentsch (1988) argued that the maximal increase in primary production should occur when the internal waves coincide with the base of the euphotic zone in eutrophic areas. In the upper part of the water column, light intensities may be great enough to inhibit photosynthesis, and internal waves propagating along a shallow thermocline in oligotrophic seas might actually decrease primary production slightly (Holloway and Denman 1989).

Internal wave packets propagate away from the shelf over deep water, as well as onto the shelf. Holligan et al. (1985) studied a situation in which a deep chlorophyll maximum layer was moved between 0.03% and 6% of surface illumination by packets consisting of 2–6 large (50–80 m) waves with a wavelength of 1–1.5 km. Some evidence indicated that mixing due to overturning of such waves could enrich the surface layer with nutrients during fortnightly spring tides.

Other vertical motions, such as Langmuir circulation, can alter the light intensity experienced over a few minutes (Marra 1980, Marra and Heineman 1982, Denman and Gargett 1983), as can the passage of clouds, and the focusing and defocusing of light by surface waves can do so in seconds (Savidge 1980, Falkowski 1984, Walsh and Legendre 1988). The issues are the depth of penetration of the effect (the spatial scale), the depth or physiological condition of the phytoplankton that makes them most responsive (Abbott et al. 1982), and the speed of change relative to the time scales of physiological change in the phytoplankton. These motions are potentially related to climatic variation through their dependence on local winds, while climatically caused changes in the density structure of the euphotic zone would be more likely to alter internal waves over continental shelves.

The issues reviewed in this section have obvious importance for the estimation of primary production and for the understanding of vertical distribution of phytoplanktonic biomass. They also illustrate the importance of scales in relations between physical processes, supplies of nutrients, and physiology of phytoplankton. Implications for the rest of the food web are less clear. In particular, it is not clear how the spatial and temporal distributions of species of phytoplankton are affected. To the extent that the functioning of the food web depends on the constituent species rather than just on biomass and primary production, the effect that vertical motions on these scales have on community composition needs further study. This observation is analogous to one I made earlier in this chapter concerning the possible role of patchiness of nutrients.

Other Effects of Internal Waves on Plankton

In addition to the physiological effects that internal waves can have on photosynthesis because of the fluctuation they cause in light intensities, internal waves can also cause phytoplankters to aggregate into horizontal patches or strips that dielly move vertically across the density interface on which the waves propagate. Under some conditions of illumination and supply of nutrients, large-celled dinoflagellates form persistently subsurface layers (described later, in Chapter 4, as a feeding ground for larval anchovy). Under other conditions, the same complex of species, when sufficiently abundant, form visible "red tides" at the surface in midday, swimming downward in the evening and upward in the morning.

Kamykowski (1979 and references therein, 1981) has modeled this effect. Figure 2.12, from his work, shows how cells, initially dispersed evenly in an onshore–offshore direction, become aggregated into bands that progress

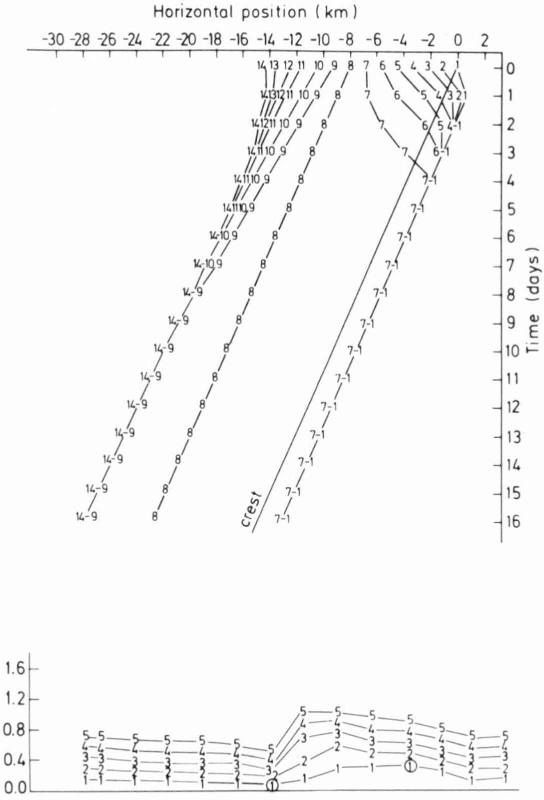

***Figure 2.12** Aggregating and physiological effects of internal waves on migrating dinoflagellates.* Upper: *offshore position of dinoflagellates on successive days in a semidiurnal internal wave. Cells, initially at positions 1–14, migrate vertically to 1 m below the thermocline each night. If thermocline is too deep for cells to reach, no such pattern is formed. Horizontal position 0 is arbitrary; negative distance increases offshore. Diagonal line shows locations of internal wave crest on successive sunsets.* Lower: *relative growth over 5 days (numbers 1–5) of cells starting at the various positions in the upper map (e.g., circles on day 1 refer to starting positions 4 and 8). Extinction coefficient k is 0.144 m^{-1} in layer above thermocline, 0.040 m^{-1} below it. From Kamykowski 1979.*

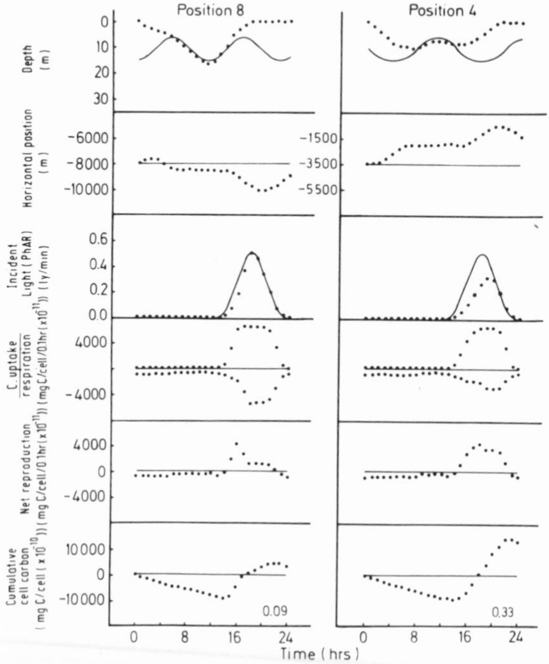

Figure 2.13 *History of cells for 24 hours, starting at positions 8 and 4 in Figure 2.12. Migration* (dotted line) *is plotted relative to thermocline* (solid line); *horizontal position* (dotted line) *relative to start* (solid line); *light experienced* (dotted line) *relative to surface light* (solid line); *and uptake, respiration, and instantaneous ("net reproduction") and integrated net accumulation of carbon. From Kamykowski 1979.*

slowly offshore over several days (as do the crests of the waves at a fixed time of day), and how the growth rates and microenvironments (Figure 2.13) of cells vary through time as a function of their initial starting position. Although the model is idealized, both in the simplification of the physical environment (e.g., no turbulence) and in the uniform behavior of the cells, it permits exploration of the importance on horizontal pattern of, for example, the depth of the thermocline (nutricline) and whether or not the cells stop swimming downward once they penetrate the nutricline at night for their "drink" of nutrients.

Zooplankton is also moved vertically by internal waves, although the effects on secondary production are less easily envisioned than the effects of the changing light field on primary production. A striking example is the work of Haury et al. (1979), who used acoustics to visualize an internal wave packet in Massachusetts Bay. A horizontal sampler, such as a zooplankton net, towed at the mean depth of the pycnocline would actually catch plankters from two distinct environments (Figure 2.14). One can imagine trying to deduce from an integrative tow at the same depth, without environmental sensors, that appendicularians and amphipods co-occur primarily in warm, relatively fresh surface water with relatively little chlorophyll.

There is an interesting relation between internal waves and recruitment

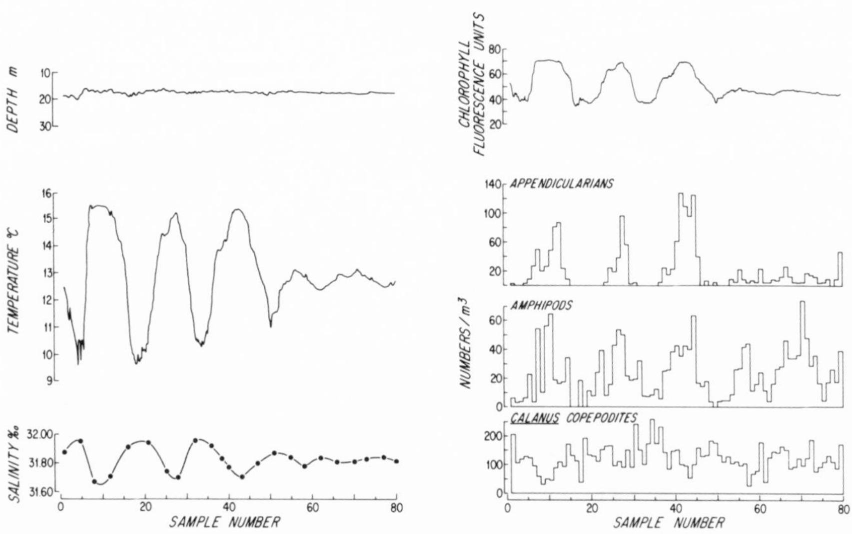

Figure 2.14 *Data from a multiple-sampling net with environmental recorders towed horizontally for 980 m through an internal wave packet: depth of tow; temperature; salinity; chlorophyll fluorescence (arbitrary units); and abundances of zooplanktonic appendicularians, amphipods, and* Calanus *copepodites. Oscillations reflect properties moving vertically across the path of the tow. From Haury et al. 1979.*

having to do with transport of larvae rather than with the dynamics of the food web. A prominent source of internal waves in stratified, shallow water is the interaction between tidal motions and bottom topography of a shelf (e.g., Haury et al. 1979); the generating mechanism is therefore relatively predictable. A sequence of linear slicks can be formed on the surface between the crest and the trough of the internal waves. These slicks are regions of weak downwelling of water, and for this reason, neustonic animals (as well as buoyant debris) become concentrated in them. Under some conditions, the slicks progress shoreward at velocities of the order of 50 cm sec^{-1}, and provide a mechanism by which neustonic larval invertebrates (e.g., the blue crab, *Callinectes)* and fish (e.g., filefish) are transported shoreward (Shanks 1983, 1988, Zeldis and Jillett 1982, Kingsford and Choat 1986).

If the recruitment of such species nearshore depends on this mechanism, climatically caused deepening of the surface layer might alter patterns of recruitment. Theoretically, one could test this hypothesis by correlating success of recruitment with the occurrences of internal waves that affect the surface (e.g., Shanks and Wright 1987, but on larger scales), but one would have to eliminate other environmental changes to demonstrate true causation.

Concluding Comment

Many of the patterns and processes I have reviewed in the latter portion of this chapter have scales too small to regulate survival and recruitment of larval fish on the scale of an entire population. I have presented some examples, related to physical and chemical processes on the same scales, of mechanisms affecting interactions between an individual and its source(s) of nutrition in the food web, and of small-scale processes that are generally homogenized or otherwise ignored in conventional measurements of rates of transfer of organic matter. These processes affect the planktonic food web by plausible, explicit mechanisms and therefore have the potential to influence the food supply of larval fish.

It is still an open question, however, as to whether significant variation in, say, success of a year class of young fish can be due to variation in these processes rather than to ones of larger scale. I shall present some evidence bearing on this question in Chapter 4, but first, in the next chapter, I turn to the mesoscale.

3
HORIZONTAL MESOSCALES AND THE PLANKTONIC FOOD WEB

The horizontal mesoscale, from tenths to tens of kilometers, is roughly intermediate between the daily ambit of a plankter and the area occupied by a population; it might be relevant to the daily movements of a school of adult pelagic fish. Operationally, it includes the range from the length of the typical tow with a zooplankton net to the interval between stations on an extensive cruise. Partly because it is accessible to conventional sampling, many investigations have been conducted on the mesoscale to define important dimensions of patchiness and relate them to the scales of physical properties.

When samples can be taken continuously, as with a fluorometer and a thermistor measuring, respectively, the fluorescence of chlorophyll in intact phytoplankters and temperature, the data can be expressed as a power spectrum, which is basically a double log plot of variance against spatial scale. The slopes of such spectra can indicate the scales at which patchiness of chlorophyll probably has a physical cause (slope similar to that of temperature, close to –5/3) and scales at which there is probably some other, presumably biological, cause. Cross-spectral coherence (i.e., significant covariation of known phase relation at particular scales) can distinguish scales at which chlorophyll varies directly with temperature from those at which it varies inversely (Denman 1976, Denman et al. 1977, Horwood 1978, Platt 1978). The brevity of this description permits only an indication of the conceptual attractiveness of this approach and conceals many difficulties in interpretation (e.g., Star and Cullen 1981).

The examples I have chosen to review here concern horizontal relations between phytoplankton and zooplankton, either as biomass or as species, together with the implications for secondary production. This issue is similar to that concerning the concordance of vertical distributions, and the implications for nutrition, discussed in Chapter 2. Continuous sampling, usually with electronic sensors, has played a major role in facilitating description of these relations (though data themselves are often sampled or averaged into finite intervals). Mackas and Boyd (1979) combined data from an electronic particle counter to estimate numbers and sizes of zooplankters near the surface along a ship's track in the North Sea and a fluorometer and thermistor to show that the zooplankton was more variable (i.e., patchier) than phytoplankton or temperature, especially on scales of 1–5 km. Cross-spectra between phytoplankton and zooplankton were coherent at 180°; that is, they were out of phase, the zooplankters being abundant where the phytoplankters were rare), which suggests grazing rather than attraction of zooplankters to phytoplankters as the cause of this scale of patchiness.

Theory (Platt and Denman 1975, Okubo 1978) suggests that the minimal size of patches that can persist in spite of diffusive dispersal increases with the generation time of the organisms of concern. Therefore, the finding that patches of zooplankton were smaller than patches of phytoplankton suggested that zooplankton patches resulted from behavioral aggregation rather than from a balance between dispersive horizontal diffusion and reproduction, since the generation time of zooplankters is normally longer than that of phytoplankters. (Actually, since it is net growth—the excess of birth over death—which must compensate for dispersion, it is theoretically possible that the grazing nearly balanced growth of phytoplankton, and so only quite large patches could persist, while the zooplankton experienced little mortality. Such an explanation is not consistent with the out-of-phase cross-correlation, however.)

After a study of somewhat similar intent, in which the zooplankters were identified and counted by microscope, Star and Mullin (1981) also reported that the intensity of patchiness of zooplankton exceeded that of phytoplankton. Although cross-spectral analyses did reveal some coherences at scales of a few kilometers in two of the three environments sampled, the coherences were not consistently out of phase.

As noted in Chapter 2, acoustic estimation of zooplanktonic biomass provides suitable data for such analysis. The variance spectrum of biomass of krill, integrated acoustically over the upper 100 m in transects in the Antarctic Ocean, was compared by Weber et al. (1986) with the variances of near-surface temperature and chlorophyll (measured as *in vivo* fluorescence) on scales of 2–100 km. The variance spectra of temperature and chlorophyll had similar slopes at the smaller scales, as indicated in other studies. Krill, which are known to form swarms by active aggregation, were much patchier at the smallest scales than either temperature or chlorophyll. Indeed, the acoustic data suggested that the actual size of individual krill swarms averaged only 13 m; unfortunately, the data on temperature and chlorophyll were recorded too infrequently to compare variances on this scale. Cross-spectral coherences indicated that variation of all three properties was in phase rather than out of phase. It is worth noting that, as Weber et al. point out, sampling at discrete stations in the same region had led to one of the earliest discussions of out-of-phase patchiness of plankton in the literature (Hardy and Gunther 1936). Other studies provide data analogous to those of Weber et al. (e.g., Pieper et al. 1990), but they have not been analyzed as variance spectra.

Implications for the planktonic food web arise conceptually from a combination of the scale-dependent coherences of distributions and the non-linear relation of ingestion to concentration of food. Huntley and Boyd (1984) and Boyd (1985) used an algorithm to convert data on horizontal distributions of phytoplankton and zooplankton in the Gulf of Maine, sampled electronically, into estimates of ingestion of organic carbon by the zooplankton and the resultant secondary production. Huntley and Boyd (1984) had concluded that concentrations of particulate organic carbon (POC) in shelf waters were generally sufficient for maximal growth, though they acknowledged that not all POC may be nutritious. However, Boyd (1985) concluded that, except during the June bloom of phytoplankton, the rate of secondary production was limited by the supply of food (as chlorophyll) at most spatial scales. Although the zooplanktonic biomass was separated into size categories and the algorithm was sensitive to this, all sizes of animals generally were either food limited or growing at maximal rates at any one time and place.

One can demonstrate that horizontal patchiness of food is significant to zooplankters by showing that some measure of physiological health, preferably one related to the rate of growth, covaries with the biomass of food on the mesoscale. For example, Hakanson (1987) found that the short-term and long-term storage lipids (triglycerides and wax esters, respectively) of the copepod *Calanus pacificus* were positively correlated with the biomass of chlorophyll in the California Current System (Figure 3.1). Because the triglycerides, as well as more conservative properties, were correlated with chlorophyll, the relations probably reflect recent nutrition, not just advection into the California Current of different crops of phytoplankton from elsewhere. Mullin (1991) demonstrated that both the production of eggs by female *C. pacificus* and the degree of food limitation of this production were positively correlated with the biomass of chlorophyll in the southern half of the region studied by Hakanson, though there was much unexplained variation.

On small scales, however, in view of the evidence given in Chapter 2, that vertical layers of food are important for *Calanus,* it is interesting that Hakanson found stronger correlations when he used the total chlorophyll averaged throughout the water column as the measure of food, rather than the maximal concentration at each station. That is, Hakanson's results do not suggest strong vertical concordance between chlorophyll and feeding by *Calanus*.

Work by Durbin and Durbin (1989) in Narragansett Bay provides other examples of horizontal variation in the condition (and, presumably, the secondary production) of zooplankton relative to biomass of food. Figure 3.2 shows the condition [as dry weight $(\text{length})^{-3}$] of female *Acartia tonsa* on a transect down the axis of the bay, and two measures of the biomass of

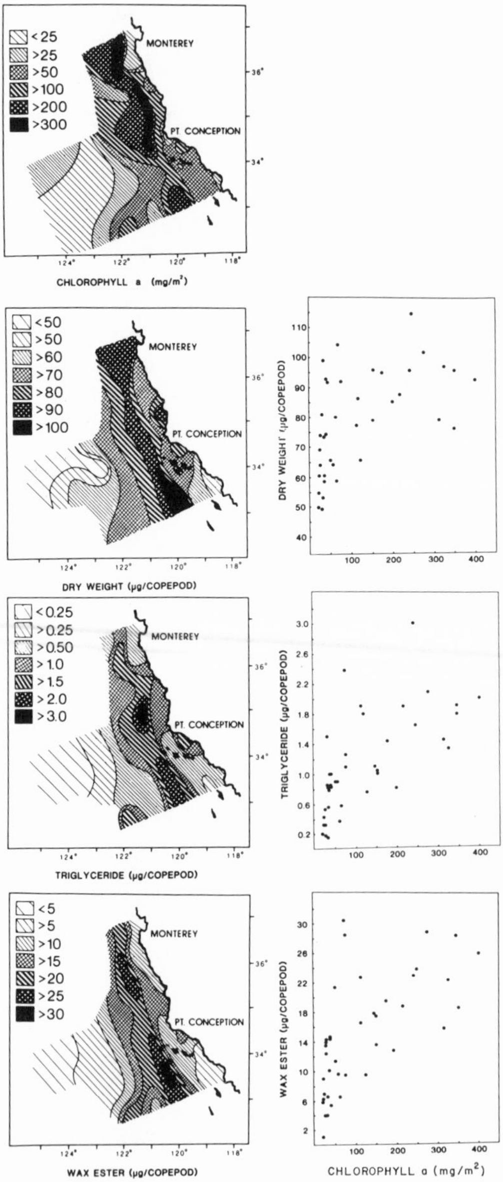

***Figure 3.1** Relationship between health and chlorophyll biomass for* Calanus pacificus *in the California Current, April 1984:* left, *distributions of chlorophyll (mg m^{-2}) and of dry weight and lipids (µg per copepod);* right, *relations of dry weight, triglyceride, and wax ester in* Calanus *to integrated chlorophyll at the same station. From Hakanson 1987.*

chlorophyll. The condition factor of the copepods appears to change more in parallel with the chlorophyll in cells larger than 10 µm than with total chlorophyll, as if only the larger cells affected the condition of the copepods.

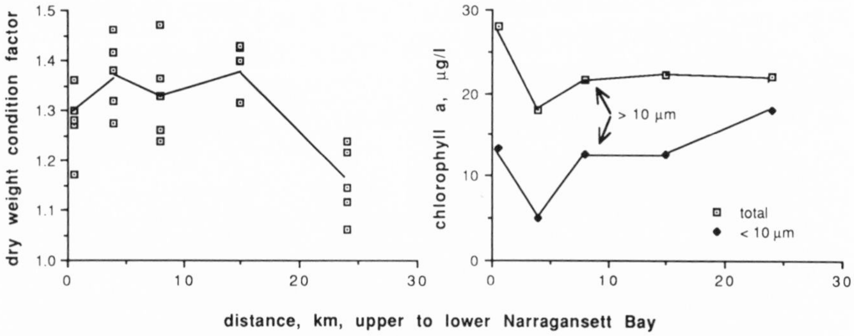

Figure 3.2 *Horizontal variation in condition of zooplankton relative to biomass of food:* left, *condition factor [dry weight (length)$^{-3}$] of female* Acartia tonsa *along a transect down Narragansett Bay, Rhode Island;* right, *biomass of total and size-fractionated phytoplankton on the same transect. From Durbin and Durbin 1989.*

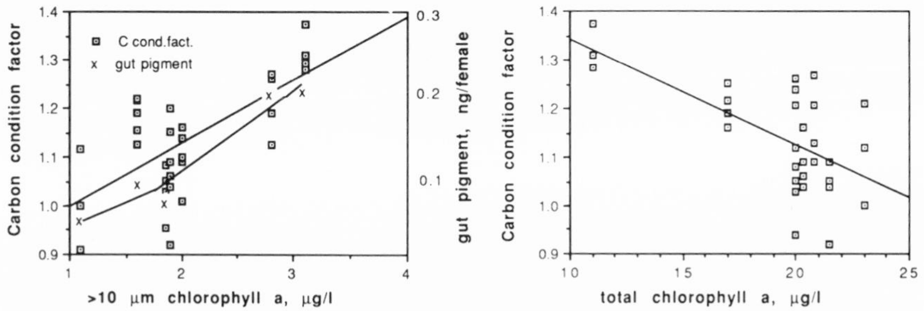

Figure 3.3 *Data taken on a 6 km transect in Narragansett Bay, Rhode Island, during a picoplankton bloom:* left, *condition factor [C (length)$^{-3}$] of* Acartia tonsa, *and phytoplankton pigments in their guts vs. biomass of phytoplankton > 10 µm;* right, *condition factor vs. biomass of all phytoplankton. From Durbin and Durbin 1989.*

This observation is supported by the results of another transect during a bloom of picoplankton (Figure 3.3); both the amount of plant pigment in the guts (as a measure of recent ingestion) and the condition factor (as C (length)$^{-3}$) increased with increasing biomass of large phytoplankters, but the condition factor was inversely related to the total chlorophyll, which was dominated by picoplankton.

The biomass of large cells is an instantaneous measurement, while condition factor is an integral of conditions of growth for the copepods over some time. On these scales, there is no assurance that the copepods experienced the observed biomass of large cells during growth, nor direct proof that ingestion of large cells caused the growth; but the data are certainly suggestive of related patterns on the scale of a few kilometers. In light of the apparent unsuitability of very small phytoplankters as food, it is worth contemplating the effect of an apparent increase in dominance of small cyanobacteria during a Californian El Niño (Putt and Prezelin 1985), mentioned in Chapter 2.

Peterson et al. (1991) sampled a transect across the Skagerrak (between Denmark and Norway) where there was variation in vertical stratification of the water column, though not a sharp front. The distribution of sizes of phytoplankters and the biomass of copepods were not clearly related to the degree of stratification, though both were variable on the mesoscale. The *per capita* production of eggs by a large copepod, *Calanus finmarchicus,* was positively correlated with the biomass of phytoplankters larger than 11 μm (which was 20%–60% of total phytoplankton), but there was no such correlation for a smaller copepod, *Paracalanus parvus.* Because growth rate of juvenile copepods was correlated with the rate of egg production by females, total secondary production by all postlarval stages of copepods could be calculated. It varied more than twofold along the transect, which was somewhat less than 100 km long, and was positively correlated (though not significantly so) with the biomass of phytoplankton. However, there is no easy way to tell how repeatable the pattern might be; that is, is it a true spatial pattern or simply variation of uncertain origin?

More extensive work in the same region by Tiselius et al. (1991) revealed similarly great variability in production of eggs by *Acartia clausi* and *Centropages typicus,* which was only partly explained by season, depth of water, and biomass of chlorophyll. The latter two factors might, of course, be related to the stability of the water column. These workers concluded that, in this region of complex mixing, the relation between measured biomass of chlorophyll and fecundity is more stable over space than over time.

Shelf Fronts

Fronts are relatively sharp interfaces between two masses of water of different physical characteristics. Some of these features are of special significance because the physical processes that establish them cause them to persist over time or (particularly for fronts resulting from the interactions of periodic currents, topography, and seasonal surface warming) to recur regularly. I shall focus on tidally generated fronts between vertically stratified and mixed water in shelf seas. Owen (1981) and Le Fevre (1986) have also provided extensive reviews of shallow-water fronts and their ecological implications.

Extensive fronts are often visible in satellite images of sea surface temperature as sharp thermal discontinuities. For this reason "interactive" cruises, in which a research vessel is guided by satellite images to sample in three dimensions, have been fruitful. Much work has been done around the British Isles on fronts generated in regions where the ratio of water depth (h) to

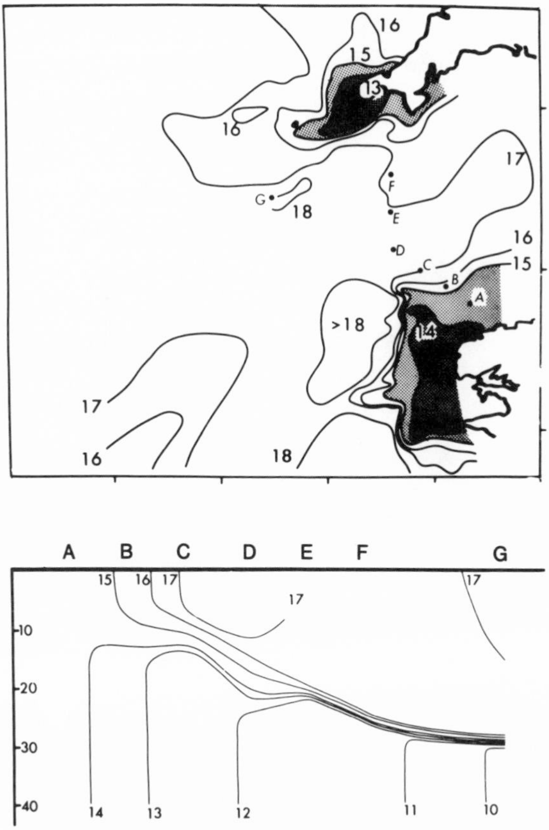

Figure 3.4 *Temperature at stations A–G in the English Channel in late July 1975:* top, *surface temperature;* bottom, *vertical section. From Pingree 1978.*

tidal current speed (u) changes rapidly. These have been called h/u^3 fronts, since this ratio sets a critical value. During spring and summer, surface heating tends to cause stratification if the value is exceeded, but mixing keeps the water column more nearly isothermal at lesser values, usually in shallower waters (Pingree 1978; Figure 3.4).

After the front has existed for some weeks, the biomass of phytoplankton (measured as chlorophyll) is often large at the surface on the mixed side of the front or at the front itself, due to the replenishment or supply of nutrients there, while there may be a subsurface layer of maximal chlorophyll on the stratified side (Figure 3.5). However, if mixing is too strong and the water too turbid, the crop of phytoplankton may be greatest just to the stratified side of the front (Pingree 1978, Fogg et al. 1985a).

Physical circulation near such a front can be quite complex, especially in three dimensions. The surface current is often parallel to the front, and there can be meanders and eddies causing cross-frontal transport (Pingree 1978, Simpson 1981, Loder and Platt 1985). Subsurface intrusions crossing the front are also thought to occur (Fogg et al. 1985b). For example, Horne et al.

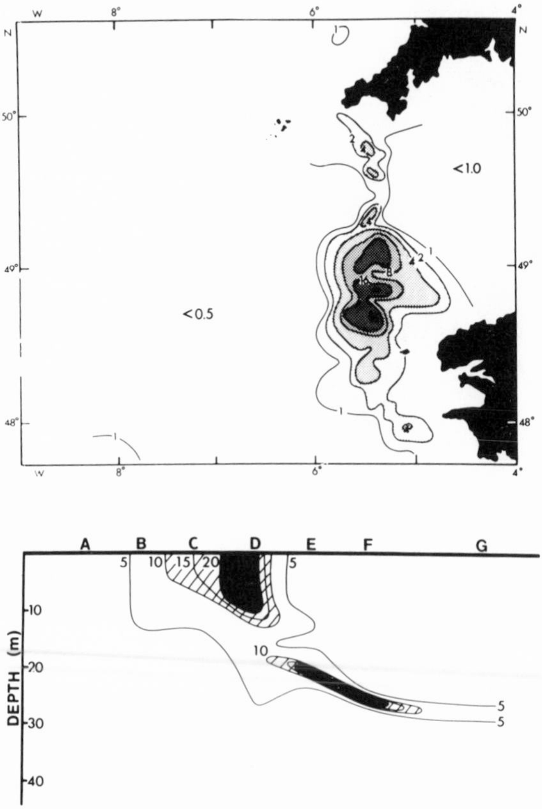

Figure 3.5 *Chlorophyll at stations A–G in the English Channel in late July 1975:* top, *surface chlorophyll;* bottom, *vertical section. From Pingree 1978.*

(1989) calculated that the supply of nitrate across a tidal front on the north side of Georges Bank from subsurface water on the stratified side was more than adequate to supply the new production (defined in Chapter 2) in the frontal and mixed regions.

An example of seasonal development is shown in Figure 3.6, and another example is described by Fogg et al. (1985b). As one might expect from the mechanism described above, variation in weather can affect the timing and magnitude of plankton blooms (Pingree 1978, Fogg et al. 1985b). Though dominant effects of present interest are in the water column, there is a relation between the critical value of the stratification parameter and the organic content of the sediment (Figure 3.7). It is not clear whether this is due to difference in production of organic matter in the overlying water column or to a direct effect of the speed of near-bottom currents on the dominant grain size of the sediment, secondarily affecting its organic content. Generally, a weaker current means finer grain size and, therefore, greater organic content due to greater surface/volume ratio of sediment particles.

Differences in taxonomic composition and in structure of the food web

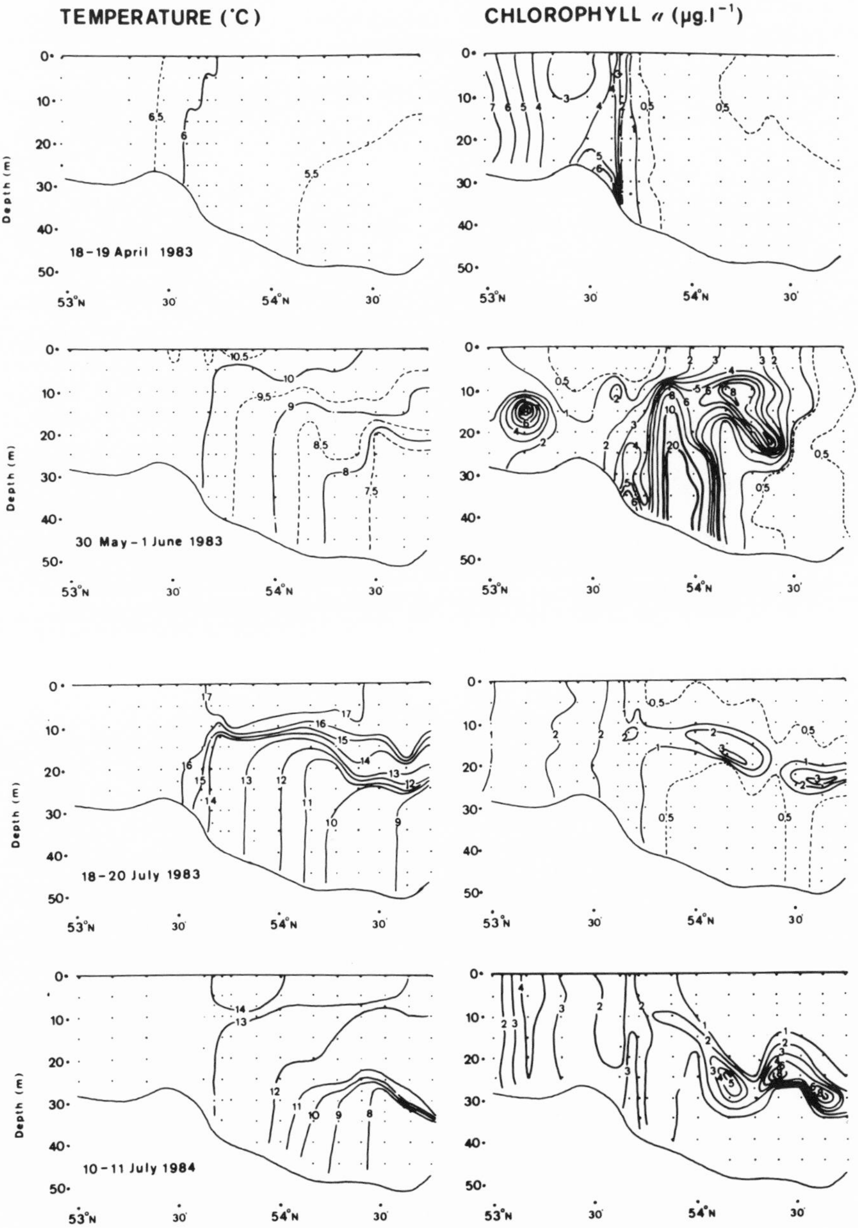

Figure 3.6 *Vertical sections of temperature in the North Sea during spring stratification (April–July 1983 and July 1984), together with chlorophyll sections. See Figure 3.7 for location. From Creutzberg 1985.*

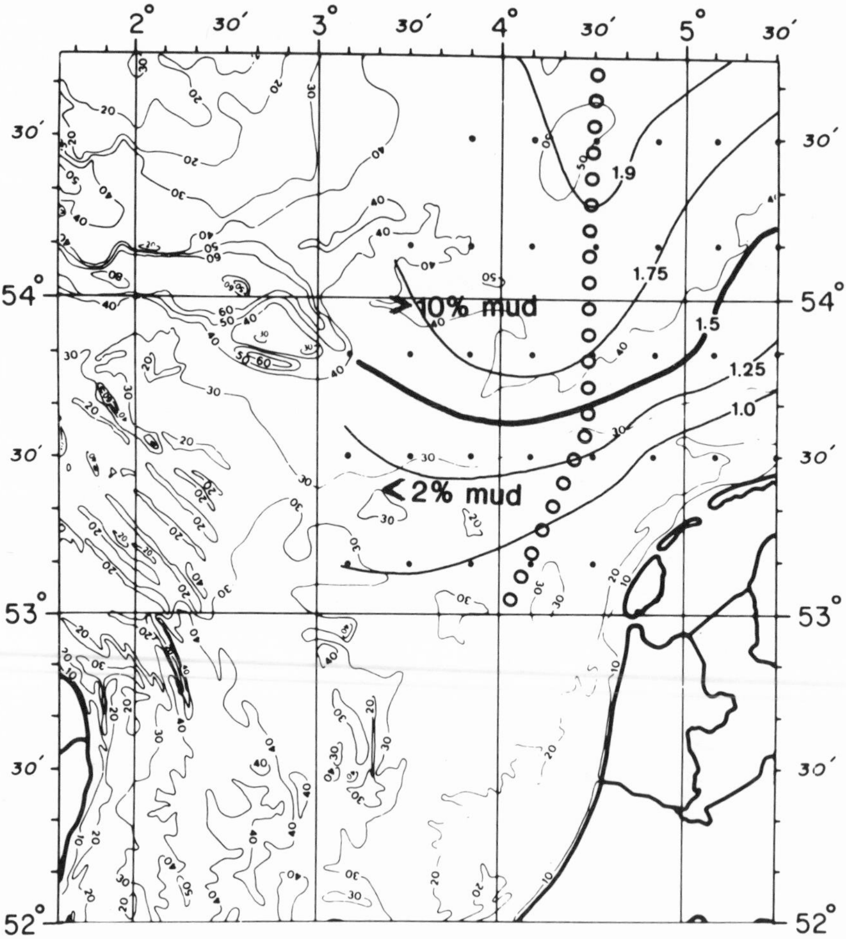

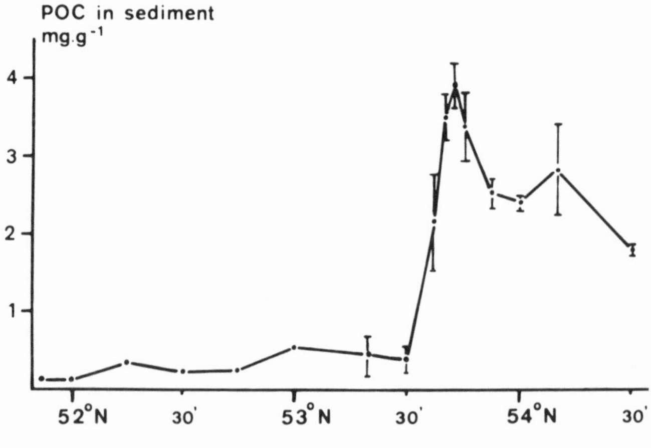

Figure 3.7 *Stratification parameter and organic sediment content in North Sea. Top: Distribution of stratification parameter (S = $\log_{10}[h(C_d u^3)^{-1}]$, where C_d is the drag coefficient and h and u are defined in text) and percentage mud (< 50 μm) in sediments, with transect line. Bottom: Particulate organic carbon in surface sediments along a transect (includes transect above). From Creutzberg 1985.*

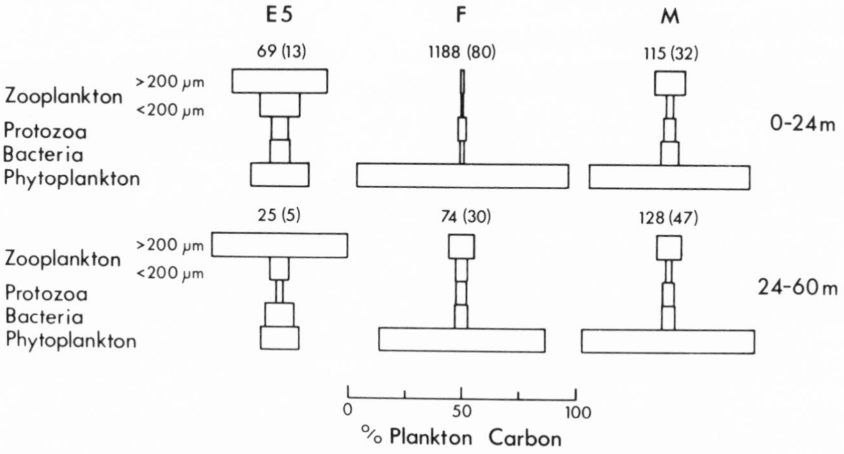

Figure 3.8 *Relative biomass (organic C) of large zooplankton, small zooplankton, bacteria, and phytoplankton in 0–24 m and 24–60 m layers at stratified (E5), frontal (F), and mixed (M) stations in the English Channel. Numbers above histograms are mean phytoplankton biomass (mg C m^{-3}) and, in parentheses, the fraction of the total particulate organic carbon attributable to living plankton. From Holligan et al. 1984a.*

may be found crossing a front, as may differences in biomass of plankton (Holligan 1981). On the well-mixed side, diatoms often dominate the phytoplankton, while dinoflagellates, such as the noxious *Gyrodinium aureolum,* may form dense blooms just across the front in stratified water (Pingree et al. 1977, Holligan et al. 1983). This pattern may be reproduced as a temporal succession of species during the springtime development of stratification on the stratified side (Holligan 1978). In at least some cases, bacterial biomass and metabolic activity are elevated in the vicinity of a front and may consume much of the primary production of organic matter (Newell and Linley 1984, Holligan et al. 1984a, 1984b, Fogg et al. 1985a), competing with the diatom-copepod-fish food chain.

Holligan et al. (1984a) presented a summary of differences in food webs between mixed, frontal, and stratified stations in the English Channel. While variability within each type of water cannot be deduced (i.e., the statistical significance of differences cannot be determined), and other fronts may present more or less contrast than did this one, Figure 3.8 is indicative of the striking differences that can be found in short distances when fronts are present, and it makes a strong argument for knowing about any frontal structure when siting stations for studies of food webs.

More important for recruitment into populations of fish than the effect of fronts on the distribution of biomass and kinds of phytoplankton is any resultant effect on zooplankton, particularly the naupliar copepods on which many larval and young juvenile fish depend for food. Holligan et al. (1984a)

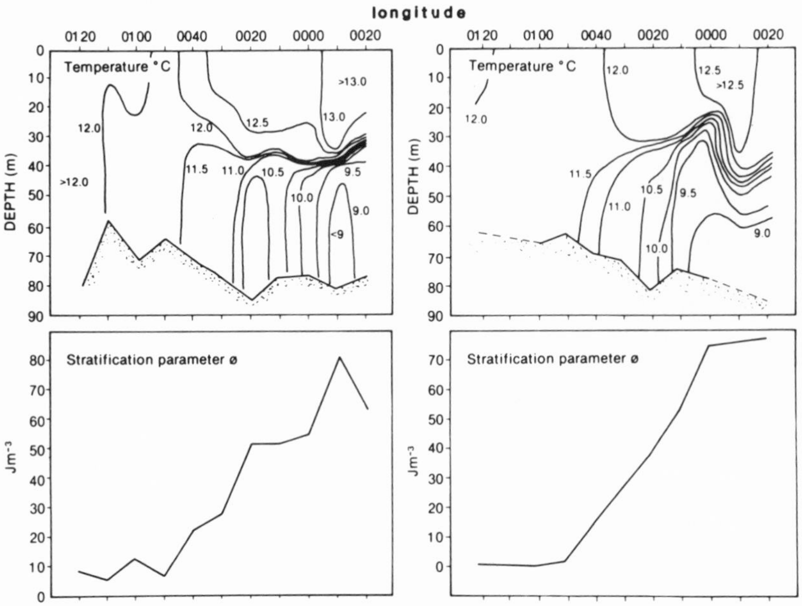

Figure 3.9 Distributions along single east–west transects of temperature and stratification parameter (a measure of the mechanical work needed to mix the column, formulated by Simpson 1981) in the North Sea east of Scotland in mid-September (left) *and late September* (right). *From Richardson et al. 1986a.*

did not detect differences in abundances of copepod nauplii at the mixed, frontal, and stratified stations they studied in the English Channel, in spite of the differences in phytoplankton (see above). However, Kiørboe and Johansen (1986) and Kiørboe et al. (1988) investigated the Buchan front on Scotland's east coast, an important nursery ground for larval herring, in two consecutive years. Their results indicate that the abundance and rate of supply of naupliar copepods sometimes varied across this front. Figures 3.9 and 3.10 show the distributions of properties in mid- and late September of 1984, the mixed water being in the nearshore, shallower region. The distribution of phytoplankton (as chlorophyll) was much patchier on the first than on the second transect in 1984 (Figure 3.10) or on two transects in 1985.

The biomass of juvenile and adult copepods was not clearly related to the front, but copepod eggs were most abundant in the mixed water in 1984 (Figure 3.11). *Per capita* production of eggs by copepods incubated on shipboard was also greatest at these stations in 1984, though in 1985 production was greatest at stations where the water column was moderately stratified. There was generally a positive relation between egg production and concentration of chlorophyll, though other factors were influential as well, since the strength of the relation varied considerably between cruises (Figure

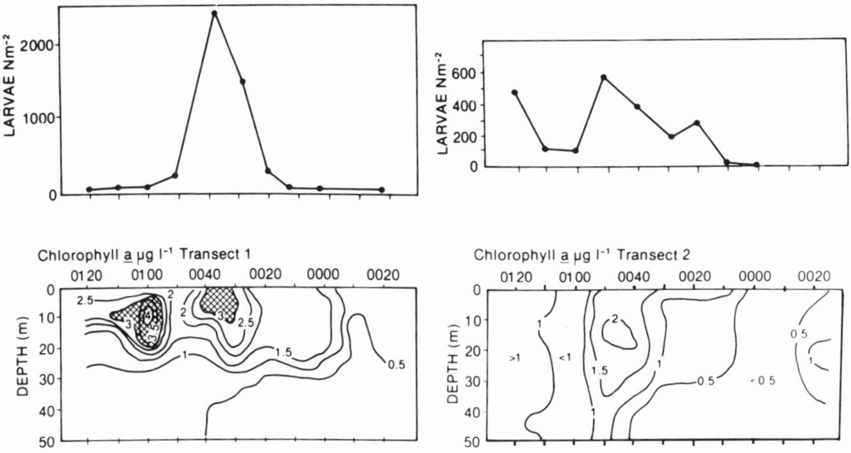

Figure 3.10 Abundances of herring larvae (top) *and distributions of chlorophyll* (bottom), *in mid-September* (left) *and late September* (right) *east of Scotland. Same location as Figure 3.9. From Richardson et al. 1986a,b.*

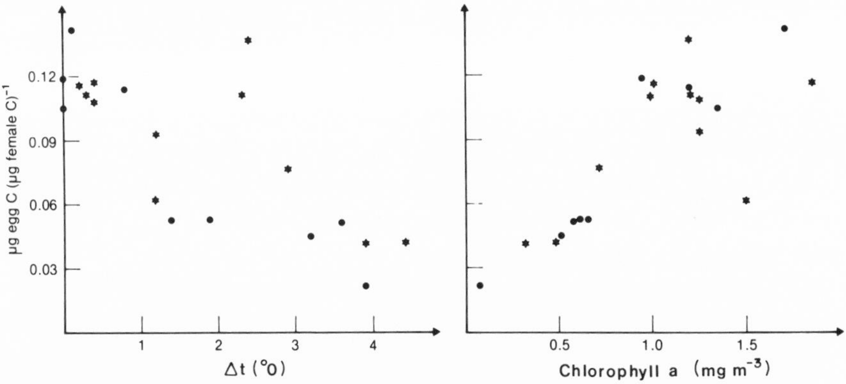

Figure 3.11 Biomass of copepod eggs per unit female biomass (sum of Pseudocalanus, Acartia, *and* Temora) *on two transects (indicated by stars and dots) in the North Sea east of Scotland, plotted against* (left) *temperature difference between 0 and 60 m (see Fig. 3.9)* *and* (right) *mean concentration of chlorophyll (see Fig. 3.10). From Kiørboe and Johansen 1986.*

3.12). Variability over time in the relation between production of eggs and chlorophyll was also shown strikingly by Tiselius et al. (1991).

The *per capita* egg production also represents, expressed as biomass per biomass, a production/biomass ratio, from which (if naupliar growth and juvenile growth are similar) total secondary production of copepods can be calculated from their biomass. On one of the two transects in 1984, this calculation showed that, with respect to the front, secondary production could be distributed differently from biomass; therefore, zooplanktonic biomass

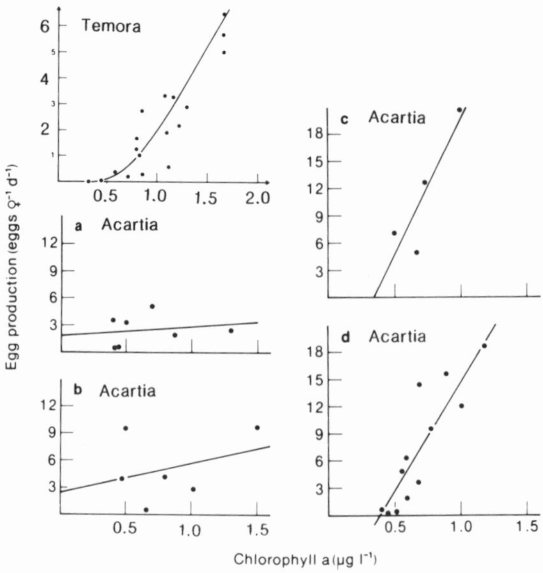

Figure 3.12 Daily egg production per female by Temora longicornis *and by* Acartia tonsa *as functions of concentration of chlorophyll where the copepods were caught on five transects in the North Sea off Scotland in two years. From Kiørboe and Johansen 1986, and Kiørboe et al. 1988.*

alone does not always indicate "where the action is" with respect to the production of fish food.

Thus, these shelf–sea fronts are potentially areas of considerable variation in the supply of food for larval fish, though the nature of this variation is itself variable in time. Quite aside from any effect mediated by the food web, fronts may also serve as regions where subpopulations of larvae are isolated and retained (Sinclair 1988).

Another type of front, related to topography but involving a different mechanism from the tidal h/u^3 fronts, often occurs at the shelf break, where a gradually inclined shelf changes to the steeper continental slope, at approximately 200 m depth. These features are important and, like tidal fronts, are quasi-constant on some spatial/temporal scales and quite variable on others. For example, Fournier et al. (1979) found that concentration of phytoplankton (as chlorophyll) was elevated at the front east of Nova Scotia, relative to well-mixed shelf waters, even though the concentrations of nutrients were similar. They attributed this to stability of the water column at the shelf break even though the stable, offshore waters beyond contained little chlorophyll. However, in a second year, the thermal front was again present but the area was mixed to greater depths and there was no augmentation of chlorophyll.

Variability in the relation between chlorophyll and temperature was also found by Marra et al. (1982) at the shelf break south of Long Island. In this case, the data permitted spectral analysis. It indicated similar (and relatively small) variability in chlorophyll and temperature at scales smaller than 1 km (though cross-spectral coherence was not calculated), but it also indicated marked variability in chlorophyll at 1–5 km due (it was argued) to growth of the phytoplankton, since variability in temperature was not unusual at this scale.

As a digression, and keeping in mind my remarks in the preface, it is worth noting that these two studies, in many ways, are typical of much of biological oceanography. Both involved considerable effort in sampling, in attempting to account for the complex physical processes affecting a dynamic region, and in considering spatial/temporal scales of various processes. Both, however, would have benefited greatly from more complete temporal coverage, with greater synopticity, such as satellites can provide (though only for near-surface properties), so that knowledge of the evolution of the patterns might (only might) clarify possible causes.

Wakes and Other Effects of Islands and Seamounts

Islands and banks in the path of currents cause mesoscale hydrodynamic effects, and in some cases these appear to influence the biomass and production of plankton. For example, Simpson et al. (1982) observed that lowered sea surface temperatures were commonly seen in satellite images around the Scilly Isles (on the shelf southwest of Cornwall), suggesting mixing of deep water to the surface by tidal currents. A hydrographic section from Cornwall into the Irish Sea indicated that the biomass of chlorophyll was greater in weakly mixed water than in the most mixed, coldest water adjacent to the Isles (however, cold surface water adjacent to Cornwall also contained a large biomass of chlorophyll). The concentration of chlorophyll was least in the most stratified water columns. Both chlorophyll and primary production in the region were quite variable, but only chlorophyll was linearly related (inversely) to the stability of the water column. Simpson et al. (1982) calculated that the islands were responsible for a sixfold enhancement of phytoplanktonic biomass over an area ten times as large as their land area.

Oceanic islands situated in a strong, steady current have been observed to generate a series of mesoscale eddies known as a von Karman vortex street. Barkley (1972), studying the phenomenon at Johnson Atoll in the mid-Pacific, reported that an eddy was shed about every two days and drifted downstream at somewhat less than the speed of the upstream current (60 cm sec^{-1}). Though the atoll is only 26 km in diameter, Barkley believed that its effect extended 600 km. The vortices are alternately cyclonic (counterclockwise in the northern hemisphere) and anticyclonic, so that doming and dipping of nutriclines should occur, as well as mixing, possibly affecting primary production (though Barkley did not study this).

At smaller current speeds, there may be trapped eddies on the downstream side of an island, but not a fully developed vortex street. Heywood et al. (1990) observed such a situation around islands in the South Equatorial Current of the Indian Ocean, together with doming of isopycnals, lowering of temperature, and enhancement of the biomass of chlorophyll near the surface. When the currents were weaker still, the effect could not be detected.

Other mechanisms can give rise to enrichment of phytoplankton around islands, of course. Dandonneau and Charpy (1983) examined data on surface chlorophyll taken from ships of opportunity in the western South Pacific, where currents are weak and the nutricline is quite deep. They concluded that statistically significant large-scale enrichment is most common well south of equatorial upwelling around high islands without extensive lagoons, rather than around all islands. Because of this, they believed that runoff of terrigenous nutrients is the likely cause of the enrichment.

Submarine banks on continental shelves have water flowing over as well as around them. Their effect on production of plankton depends on their size and the characteristics of the water reaching them, and on the degree of vertical mixing and "trapping" of water over the bank in a gyral circulation relative to the rate at which water leaves the bank (Loder et al. 1988). Figure 3.13 is a sketch of the relative magnitudes of vertical diffusion, recirculation, and residence on four banks in the western North Atlantic: Flemish Cap (deep) and Southwest Shoal of the Grand Banks (shallow) off Newfoundland, and Browns Bank and Georges Bank surrounding the Gulf of Maine (both are strongly impacted by tides). Georges Bank has both the absolute size and the relative importance of vertical diffusion to have a great effect on primary production. Loder et al. (1988) suggest that scaling the hydrodynamic processes in this way is essential to compare the effects of the banks with biological time scales.

Oceanic seamounts (which rise from water more than 1 km deep) are capable, under certain conditions of current and stratification of the water, of setting up an anticyclonic vortex (a Taylor column) that can remain trapped above the seamount for weeks, causing distortion of isotherms and local enrichment. Genin and Boehlert (1985) observed uplifting of isotherms (which is possible because the anticyclonic rotation is not in geostrophic balance) and vertical profiles of chlorophyll above Minami-kasuga Seamount that were different from those at "control" stations away from it (Figures 3.14 and 3.15). However, they did not detect these phenomena on two other surveys.

The large populations of demersal fish that often inhabit such seamounts may obtain part of their nutrition from macrozooplankton and micronekton

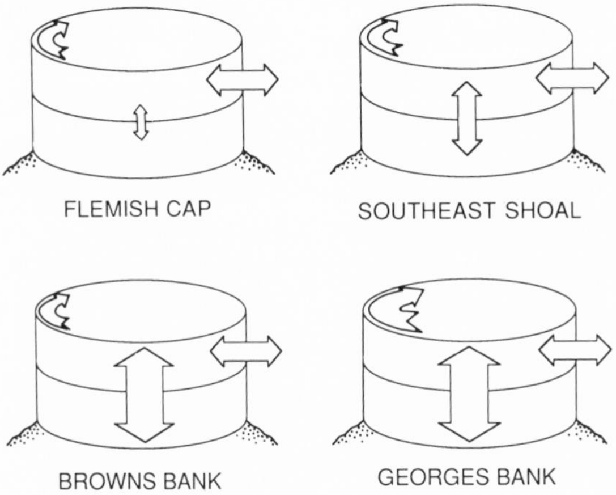

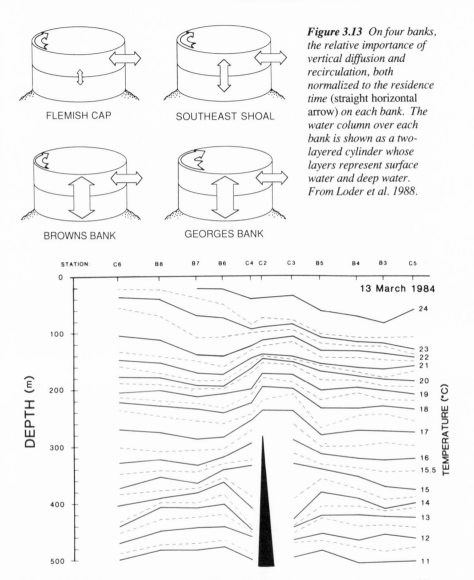

Figure 3.13 *On four banks, the relative importance of vertical diffusion and recirculation, both normalized to the residence time* (straight horizontal arrow) *on each bank. The water column over each bank is shown as a two-layered cylinder whose layers represent surface water and deep water. From Loder et al. 1988.*

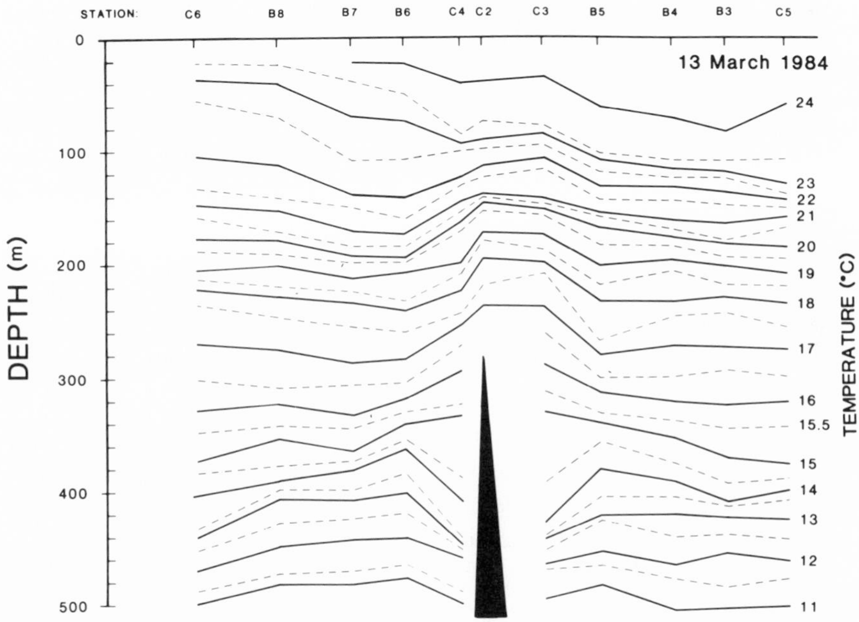

Figure 3.14 *Isotherms around and over Minami-kasuga Seamount. The transect was 22 km long; note vertical exaggeration. From Genin and Bohlert 1985.*

whose auroral downward migration intercepts the seamount (Isaacs and Schwartzlose 1965). Genin et al. (1988) reported evidence supporting the hypothesis that the distribution of migrating zooplankton becomes patchier in the vicinity of rough topography, but they attributed this effect as much to the hydrodynamic effects of the topography as to predation by demersal fish.

These examples illustrate the effects of topography in creating meso-scale variation both on continental shelves (banks, islands) and in the open

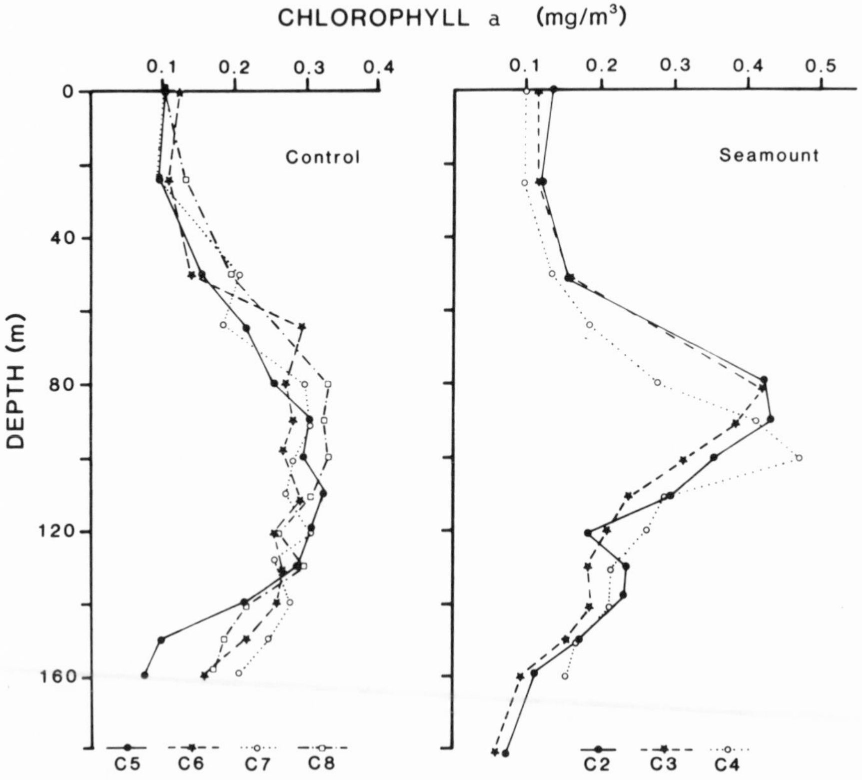

Figure 3.15 Vertical profiles of chlorophyll at "control" stations (left) *away from the seamount and* (right) *over the seamount shown in Figure 3.14. From Genin and Bohlert 1985.*

ocean (seamounts, islands). There is less direct evidence of the effect on larval and juvenile pelagic fish, though the distribution of some pelagic fisheries suggests that adult populations are associated with topographically caused mesoscale features (Uda and Ushino 1958).

Miller (1974), working off Maui in the Hawaiian Islands, reported evidence that concentrations of larval fish and zooplankton were elevated on the upstream face of a small islet (Molokini) while current was flowing, but too few samples were taken to establish the generality of this result. The larger Hawaiian Islands often have eddies of 50–60 km diameter in their vicinity. Lobel and Robinson (1986) argued that the overall effect of such eddies on the reef populations is a balance between the removal of larvae from the reefs by entrainment and the retention of such larvae in the immediate vicinity of the island in the eddy for up to two months. Effects on larval fish of eddies of other origins are discussed in the next chapter.

Oceanic Eddies

Textbooks frequently depict an oceanic boundary current (such as the Gulf Stream, the Chile–Peru Current, the Kuroshio Current off Japan, and the Benguela Current off West Africa) as smooth, straight flows separating waters of the continental shelf and slope (relatively cool and fresh) from oceanic central water masses. While this correctly represents the mean flow, such a current usually consists of a series of meanders or loops. When one of these loops becomes sufficiently exaggerated, a rotating eddy of a few hundred kilometers in diameter is pinched off and can drift away from the current.

In the North Atlantic, an idealized "cold-core" eddy consists of a ring of Gulf Stream water with cyclonic (counter-clockwise) rotation located east and/or south of the Gulf Stream, surrounded by the relatively warm, oligotrophic Sargasso Sea, and enclosing relatively cool, eutrophic water from the American slope or shelf. A warm-core eddy is the opposite: a ring with warmer water at its center and anticyclonic rotation, drifting west and south of the Gulf Stream along the continental slope.

The surface temperatures and (sometimes) the biomasses of phytoplankton of these rings differ from the surrounding water, creating quasi-circular oceanic fronts that can be detected and tracked by satellites (e.g., Gordon et al. 1982). There may be a relation to bottom topography, since seamounts off New England may influence the formation of the Gulf Stream meanders (Richardson 1983). Because they represent an isolated, large example of one environment and biotic community "invading" another (Wiebe et al. 1976), and because a large fraction of the ocean's kinetic energy is in such mesoscale features (e.g., Richardson 1983), the rings have been the object of intensive study over the past fifteen years.

At any given time there are likely to be approximately ten cold-core rings in the Sargasso Sea, covering about 10% of its surface area, and ranging up to a year or more in age since separating from the Gulf Stream (Ortner et al. 1978). Current velocities can reach 1.5 m sec^{-1} in the region of fastest rotation of a young ring, and the features tend to drift to the southwest through the Sargasso Sea at mean speeds of 5 cm sec^{-1} (Richardson 1983). The surface manifestation of the rings may be lost due to local warming and mixing with Sargasso Sea water while there are still remnant anomalies between 200 and 800 m (Wiebe et al. 1976). While the ring is aging, its fauna may change from that characteristic of the continental slope to that characteristic of the Sargasso Sea (Wiebe et al. 1976), but for phytoplanktonic species (which have a short generation time), conditions may permit development of a flora that for a time is not simply intermediate between the two (Ortner et al. 1979).

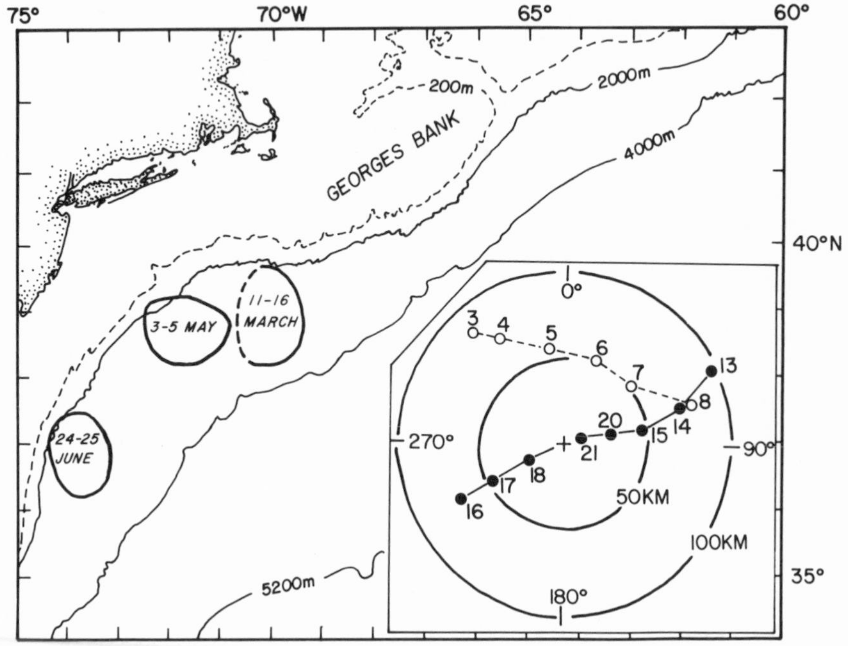

Figure 3.16 *Positions of warm-core ring 82-B off the Middle Atlantic states in March, May, and June 1982. The continental shelf is shallower than 200 m; the steepest part of the slope is between 200 and 2,000 m. Inset shows transects resulting in Figures 3.18, 3.19, and 3.20. From Nelson et al. 1985.*

As a ring ages, the populations it contains may become subtly unlike the slope-dwelling populations from which they were derived. Euphausiids *(Nematoscelis megalops)* live deeper in the water column in a cold-core ring than over the slope. Individuals from a ring that is nine months old contain less lipid and less organic carbon and nitrogen, respire at lesser rates, and generally show more signs of physiological stress than do individuals from slope water (Wiebe and Boyd 1978, Boyd et al. 1978).

Approximately five warm-core rings are formed from the Gulf Stream each year. They seem to change more rapidly than cold-core rings. They tend to lose their surface identity by contact with the atmosphere, because they drift at a few centimeters per second to the southwest (Figure 3.16), where the Gulf Stream lies close to the American shelf. They are also likely to be altered by contact with the shelf and slope (since they are as much as 1 km deep) or by interaction with the Gulf Stream (e.g., Evans et al. 1985). A common fate of such rings is to re-enter the general flow of the Gulf Stream within about six months.

Although warm-core ring 82-B, formed in February 1982, was probably studied as intensively from a biological perspective as any oceanic mesoscale feature has been, changes due to ecological interactions within the ring were not always easy to separate from those due to intrusion of surrounding water from various sources. Phytoplankton and other seston decreased in surface waters due to deep convective mixing in March 1982, as well as to grazing (Bishop et al. 1986). Vertical structure within a ring can also be disrupted by storms, as seen in ring 81-D in September of 1981 (Hitchcock et al. 1987). When sampled in April 1982, the phytoplanktonic biomass of ring 82-B was rather uniform with depth, and much lower than in the surrounding slope water. By June, however, the upper 100 m of the ring was stratified, and the phytoplanktonic biomass (especially siliceous forms, indicated by biogenic silica) equaled or exceeded that in the slope water (Figure 3.17). Therefore, the ring was not a permanently oligotrophic "desert" surrounded by rich slope water but a mesoscale feature whose spring bloom of phytoplankton had been delayed (Hitchcock et al. 1985).

The source of nutrients for this bloom was a counterintuitive physical process. Since warm-core rings are anticyclonic, downwelling of water within the ring is expected, tending to move nutrients away from the surface. Warm-core rings decay or wind down through friction, however, and this causes upward movement of the deep pycnocline and nutricline, aiding (but not entirely accounting for) the supply of nutrients for "new" production (as defined in Chapter 2) (Franks et al. 1986, Nelson et al. 1989).

Transects across ring 82-B in June (see inset, Figure 3.16) show more clearly the mesoscale distribution of properties within and around the ring (Figure 3.18). The steep downward sloping of the isotherms shows the anticyclonic flow at the edges of the ring. On the second transect, a secondary feature called a shelf-water "streamer" was encountered about 75 km to the northeast of the ring's center. Figures 3.19 and 3.20 show the distribution of biological properties on this second transect. The concentrations of chlorophyll and particulate ATP (a measure of metabolizing biomass) were greatest at a depth of 25 m around the center of the ring and in the streamer, while particulate organic carbon, measured either directly or by counting particles electronically ("Coulter POC"), detrital carbon (total minus living POC), and biogenic silica were most concentrated at the center of the ring. Bacterial biomass and productivity, however, were greater towards the edges than at the center, as was the biomass of picoplankton (Peele et al. 1985, Ducklow 1986).

Within rings, the subsurface maxima of chlorophyll and ATP (Figure 3.19) may be maintained in part by the distributions of grazers. Roman et al. (1986) reported that small zooplankters were less abundant in these maxima than at shallower depths.

The zooplankton of ring 82-B was studied by size class, since different

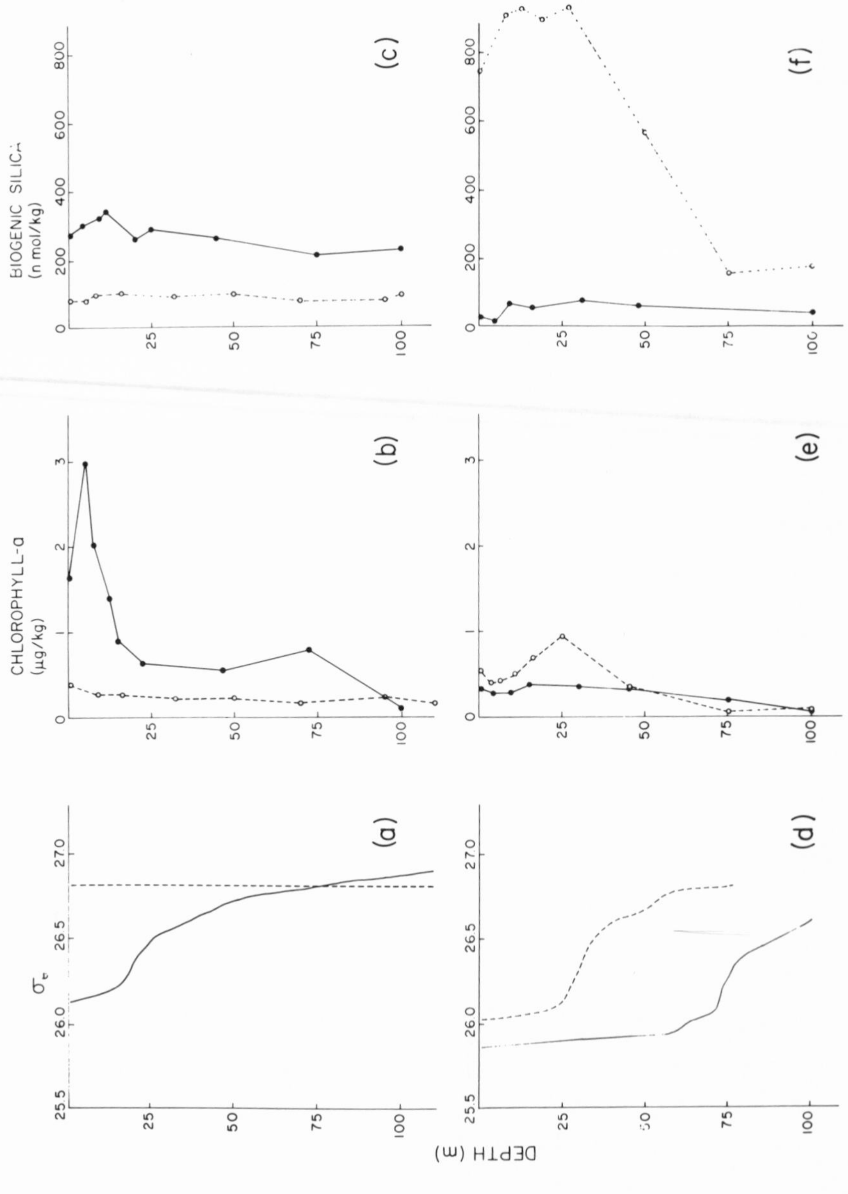

Figure 3.17 *Vertical profiles of density, chlorophyll, and biogenic silica in the center of ring 82-B* (dashed line) *and slope water* (solid line) *in* April (a, b, c) *and June* (d, e, f) *at location shown in Figure 3.16. From Nelson et al. 1985.*

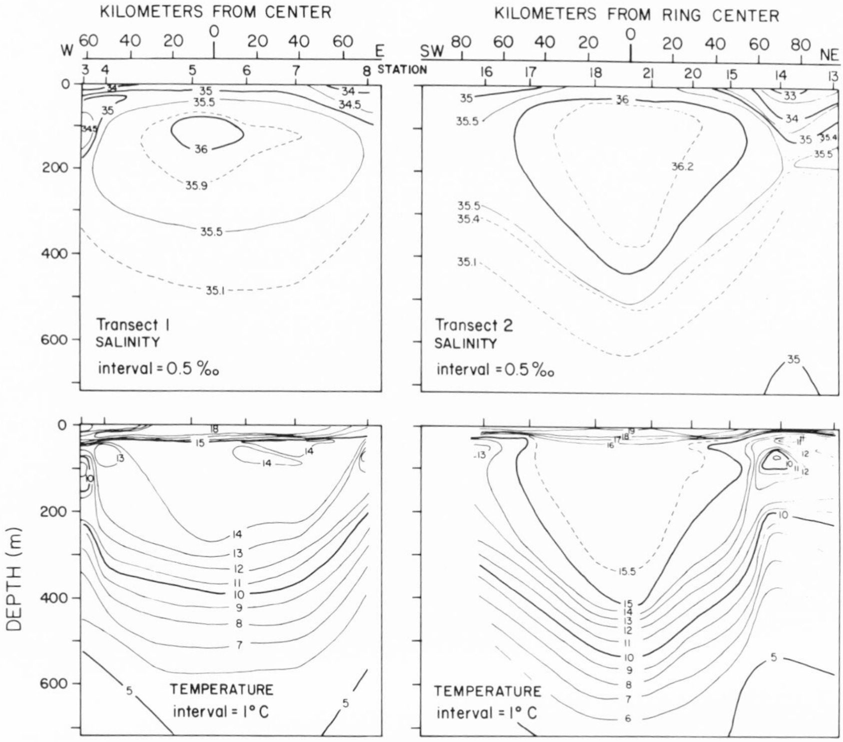

Figure 3.18 Vertical sections of salinity (upper) *and temperature* (lower) *through ring 82-B. See Figure 3.16 for transects 1 and 2. From Nelson et al. 1985.*

meshes of nets filtering different volumes of water were necessary. The biomass of macrozooplankton (> 500 μm) was greater in the ring in June than in April, and the increase was primarily due to increased abundance of larger (possibly carnivorous) zooplankters such as euphausiids (Davis and Wiebe 1985). The estimated consumption of macrozooplankton by micronekton was sufficiently small to be consistent with the increase in the former (Boyd et al. 1986). Roman et al. (1985) did not note any difference in the biomass of microzooplankton (64–333 μm) between April and June, nor did they record an increase in zooplankters larger than 333 μm when they sampled with a different net. That any increase in biomass occurred in the largest types of zooplankton was somewhat surprising, since the late bloom of phytoplankton in the ring should have stimulated the growth of smaller, herbivorous copepods instead.

By June, the biomass of macrozooplankton in the ring exceeded that in slope water and was concentrated at shallower depths (Wiebe et al. 1985). Microzooplanktonic biomass in the ring also exceeded that in slope water in June, but this was because the biomass in slope water was less in June than in

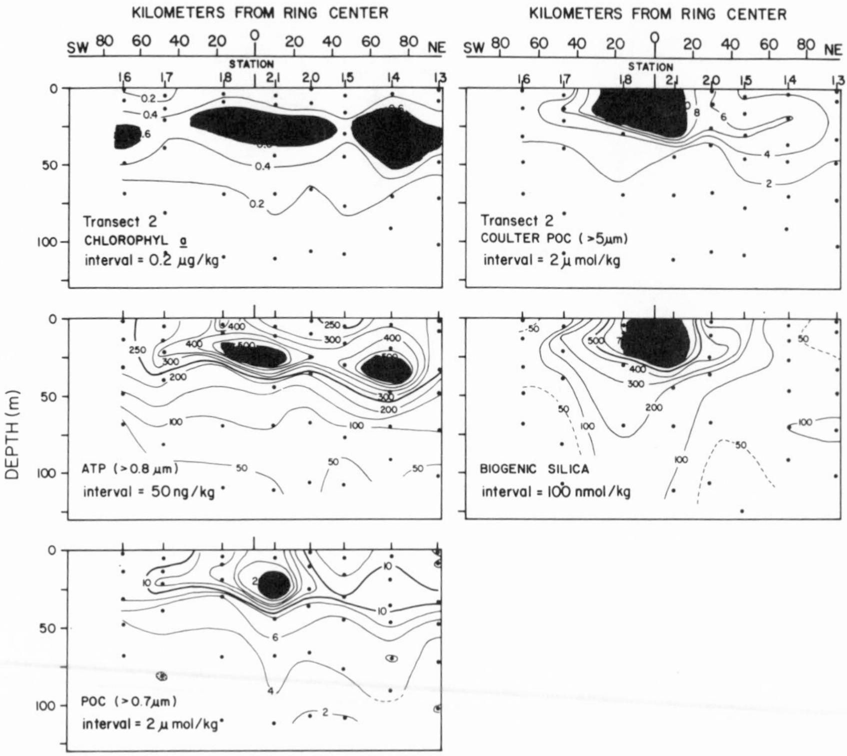

Figure 3.19 *Vertical sections from transect 2 (see Figure 3.16) of chlorophyll (shading indicates > 0.6 µg l⁻¹); organic carbon in particles > 5 µm, estimated by electronic counting (shading > 10 µM); particulate ATP (shading > 500 ng l⁻¹); biogenic silica (shading > 700 nM); and particulate organic C (shading > 20 µM). From Nelson et al. 1985.*

April, not because biomass in the ring had increased (Roman et al. 1985). Off Japan, Yamamoto and Nishizawa (1986) reported a marked increase in abundance of several species of zooplankton crossing from slope water into a warm-core ring of the Kuroshio Current.

The studies of ring 82-B, based on cruises several weeks apart, again illustrate the difficulty in distinguishing changes due to food web processes within the ring from changes due to advection and mixing. (Perhaps as a result of this experience, some of the investigators are attempting to develop an acoustic buoy that can record zooplanktonic biomass while drifting untended.) At least five hydrographic intrusions occurred during the life of the ring, and it was in the process of being absorbed in (or overwashed by) the Gulf Stream when it was studied in August. The "streamer" (Figure 3.18) had particularly high concentrations of microzooplankton, which might have affected the biomass in the ring. Wiebe et al. (1985), who attempted to distinguish *in situ*

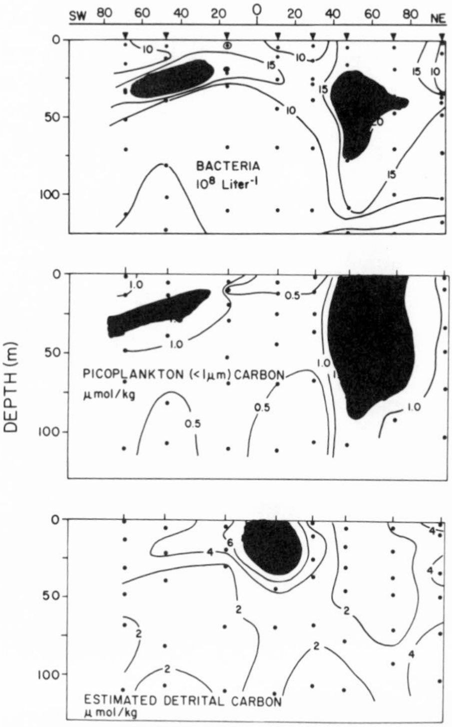

Figure 3.20 *Vertical sections from transect 2 (see Figure 3.16) of bacterial cells (shading > $2{\cdot}10^7\ l^{-1}$); picoplankton (shading > 1.5 µM); and detrital organic carbon (shading > 8 µM). From Nelson et al. 1985.*

from advective changes by using changes in salinity to indicate the advective changes, concluded that much of the increase in macrozooplanktonic biomass was due to growth *in situ*. Partitioning the increase among species of known biogeographic affinity (i.e., slope water versus Gulf Stream species) might have clarified matters.

In general, storm-generated turbulence could disrupt a ring, as it is on the right scale. Such disruption has been difficult to observe, however. Cowles et al. (1987) commented on changes in zooplanktonic biomass in ring 81-D after a storm and on the implications for grazing on phytoplankton, but the sampling was insufficiently replicated (in my opinion) to trust their conclusions.

In the California Current, there is a region approximately 150–250 km offshore in which mesoscale eddies are particularly common, and may be important in moving coastal populations offshore and oceanic populations shoreward (Simpson 1987). Study of one such warm-core eddy revealed processes analogous to those in Gulf Stream rings, but this eddy seems to be persistent, or at least to recur in more or less the same location, in several years (Simpson et al. 1986).

The eddy was detected and followed by satellite imagery, but its vertical structure was quite complex, and so the relations between the vertical distributions of chlorophyll and temperature and the data gathered by CZCS and AVHRR sensors (designed to measure these properties near the surface) were quite different from the relations in the surrounding water. Entrainment of cold water containing a large biomass of chlorophyll, probably originating many kilometers northeastward as coastal upwelling, was noted, as was alteration of structure by a windstorm. Analysis of species of zooplankton occurring in and around the eddy also suggested entrainment of water from several sources. The species were stirred together (though not mixed to homogeneous composition in the eddy), but the overall faunal composition was not unusual (Haury 1984, Haury et al. 1986).

In summary, rings and eddies can provide mesoscale sites in which conditions are quite different from surrounding water, even though they have proved less isolated from their surroundings than originally expected. One consequence is that a time series of data designed to study seasonality (e.g., Menzel and Ryther 1960) can be distorted by an eddy that crosses the sampled area but is not recognized as such. The food web may have a different structure, different timing, or both, compared with the surrounding water. In addition to their potential advection of larval fish, eddies can alter migration patterns of adults. Saitoh et al. (1986), in comparing the distribution of commercial catches of saury in their southward migration from the Oyashio to the Kuroshio off Japan with AVHRR images of sea surface temperature, concluded that schools avoid and migrate around mesoscale eddies. It would be interesting to know whether the fish in that study were responding to the physical or to the biotic conditions at the fronts around the eddies.

Rhetorical Questions and Comments

In this and the preceding chapter, I have illustrated by some examples the relations between physical processes on micro- and mesoscales and the distribution and productivity of plankton on these scales. For brevity, I have omitted other, equally good examples, the most notable being coastal upwelling (see, e.g., Barber and Smith 1981). To the extent that planktonic productivity generally, or the production of specific categories of plankton, constitutes the supply of food for larval and juvenile fish, recruitment to fish populations can be affected (specific cases are presented in Chapter 4). However, I believe that the fraction of interannual or spatial variability in fish recruitment that can be accounted for statistically by variation in planktonic production is not known for any pelagic population, and so we are proceeding based on a qualitative, though reasonable, belief. In fact, until better methods are devised for estimating the secondary production of those plankters actually used as food by young fish, we are unlikely to improve the situation. I will return to this issue in the final chapter.

The relevance of Chapters 2 and 3 depends on the validity of the assumption that to understand fully the organization and dynamics of pelagic ecosystems, so as to predict how they will respond to very large-scale climatic change, one must first understand the physiology, behavior, and ecology of plankton on scales much smaller than an ecosystem. It is useful to speculate on some possible changes in physical processes on these smaller scales.

Tidal fronts are related to the seasonal development of stratification on one side of a boundary that is determined by depth and current speed. A presumed mechanism for the ecological importance of such fronts is the close proximity in space and time of nutrient enrichment and a stable water column. Will this proximity change in response to climatic warming, e.g., because of an altered relation between nutrients and density? If so, will fronts be less of a "hot spot" for secondary production and hence for growth of larval fish? If Sinclair (1988) is correct that the importance of fronts, at least for herring, is due more to the physical retention of larvae within a genetic subpopulation than to an augmented supply of food, the effect of climatic warming on fish recruitment will be modest, even if primary and secondary productivity are altered, as long as the hydrographic properties of the front are retained.

I reviewed the effects of mixing and internal waves on primary production. Given that climatic warming and increased stratification of surface waters is likely to be most intense toward the poles (especially the North Pole, which lacks the South Pole's vertical circulation to carry away heat), is it likely that mixing will become less important there and internal waves more so? If it is, will floristic composition, vertical distribution, or patch structure change as a result of this shift, or will the limiting factor for primary production in high latitudes change from light to nutrients? If the pycnocline is deeper, will internal waves therefore be less likely to provide a mechanism of shoreward transport for neustonic larvae?

Figure 3.21 shows a relation between characteristic patch size, patch intensity (as determined by the coefficient of dispersion), and mean abundance of developmental stages of a common copepod. Will this relation apply to decreased mean abundances over a large area, and will the foraging strategies of zooplanktivorous fish need to change in emphasis from finding and remaining with a dense patch to rapid transit between less intense patches?

Given that there is mesoscale variation in primary and secondary production, how important is this in the longer term? Is variability within a

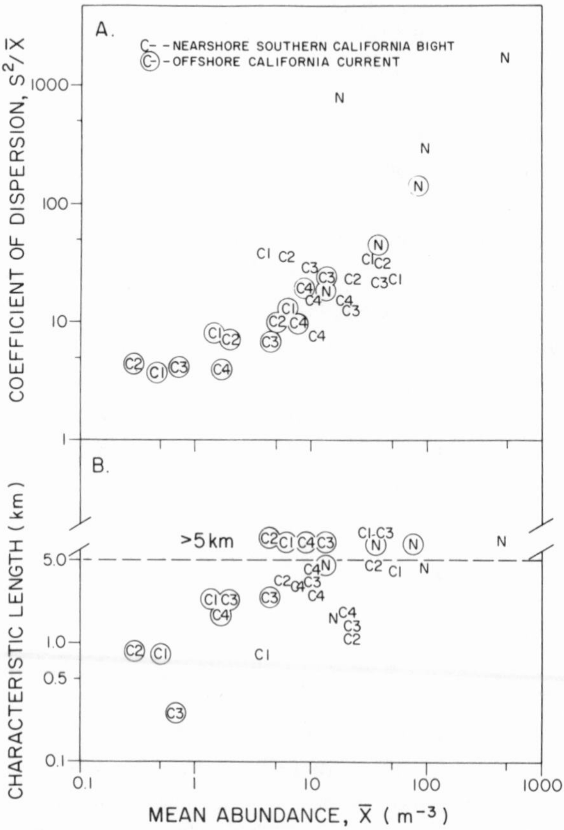

Figure 3.21 *Coefficients of dispersion (variance* x *mean*$^{-1}$*) and characteristic length scales (km) of distributions of* Calanus pacificus *developmental stages (N = all nauplii, C1–4 = copepodite stages I–IV) as functions of mean abundances on diurnal and nocturnal horizontal transects in the California Current (circled) and nearshore portion of the Southern California Bight. The exact values for length scales > 5 km could not be reliably calculated. From Mullin 1988.*

population of zooplankters in lifetime fecundity, which integrates over time, greater or less than variability in gut contents, RNA/DNA ratio, or daily egg production, and how much genetic variation is there? Under what conditions, or for which populations, does successful recruitment depend on satisfactory median conditions for juvenile stages over most of the reproductive range? For which populations must a small but important percentage of larvae co-occur with anomalously high concentrations of food for a considerable part of larval life? Consider, as examples, Figures 3.1, 3.3, and 3.12. Are the mortality rates of the respective populations such that *Calanus* with > 20 μg wax ester, or *Acartia* with carbon condition factor > 1.2, or *Temora* producing > 5 eggs per day, must be present somewhere in each population's range for the population to survive?

These issues and similar speculations are intellectually stimulating, but until the small-scale effects of climate change are better understood, they probably cannot be resolved. There are, however, several areas in which I believe progress is needed soon.

It is essential to embed more thoroughly and more frequently small-scale measurements within studies on larger scale, with the intent to determine whether increase in understanding actually results (assuming that understanding of variation on scales of populations and communities is the ultimate goal). For example, one could combine analyses of gut contents of zooplankters with thermistor microprobe detection of "patches" of turbulence to see whether the animals are, in fact, more efficient at utilizing the phytoplankton in such patches, as the model by Rothschild and Osborn (1988) would suggest. One might also try to determine whether the rates of growth of populations in the zooplankton show some spatial or temporal correlation with the intensity of turbulence on a large scale.

Even if those two studies yielded positive results, one still needs to perform the analysis of gut contents and turbulence on small scales in the context of a larger study, to reduce the possibility that some covariate of turbulence, or some ecological interaction other than feeding on phytoplankton, actually caused the variation in growth rates of the zooplankton. In formal terms, if one has a favorite, particular, small-scale explanation of the large-scale phenomenon, one must not only obtain data consistent with it but also try to eliminate other possible explanations. In less formal terms, those of us who do this work need to be able to move the topic of small-scale causes and large-scale effects from "Discussion" to "Results" in our published papers.

The amount of work required to analyze the variety of small-scale interactions in terms of their real roles in the dynamics of a single population or a whole community renders this goal virtually unattainable, at least with present tools. Are we really going to be able to analyze the mesoscale population genetics of enough pelagic species (e.g., Bucklin et al. 1989) to determine which mesoscale variation in population dynamics is genotypic and which ecophenotypic? On the other hand, such practical impediments do not mean that the goal, as a goal, is meaningless.

We need to translate into specific terms for pelagic ecosystems the theory that crucial, small-scale relations exist between the physical/chemical environment and the populations that will determine the system's response to large-scale perturbation, and to identify the relations. As I noted in Chapter 1, there is theoretical reason to think that such crucial relations exist, but for most ecologists who work on small scales, this is (I suspect) simply an article of faith or a rationalization for choosing tractable, well-encompassed problems. There is obviously a strong role to be played by further development of the theory of functioning ecosystems, but my sense is that much present theoretical work is dominated by "terrestrial" (and intertidal) paradigms, where competition for living space as such is intense.

Further (and this also returns to a theme of Chapter 1), we need to be able to measure abundance and rate of production on many scales, and to compare the scale-dependent variability with that of physical and chemical properties. Variance spectra and cross-spectral analysis are one means of accomplishing this, and we need to employ more fully their underlying philosophy of measurement, if not their computational formalism, particularly for the rate processes contributing to production. This requires new technological tools, and more widespread availability of the tools that already exist, but it also requires shrewdness in the design of sampling programs, or in drawing new insights from programs designed for other purposes.

4
LARVAL ECOLOGY AND RECRUITMENT OF PELAGIC FISH IN THE CALIFORNIA CURRENT

Up to this point, I have tried to describe ecological processes (and the underlying hydrodynamics) rather than specific environments or seas. In this chapter, I reverse the emphasis. Investigations of the ecology of commercially important pelagic fish in the California Current System—the anchovy *(Engraulis mordax),* sardine *(Sardinops sagax),* chub or Pacific mackerel *(Scomber japonicus),* jack mackerel *(Trachurus symmetricus),* and whiting or hake *(Merluccius productus)*—illustrate the range of scales and types of investigations through which understanding of variability in populations has been sought. The physical and chemical environment these species occupy has been thoroughly studied, as have the biomasses of zooplankton and (more recently) phytoplankton. Much of the work was conducted by collaboration between oceanographers and fisheries scientists in the California Cooperative Oceanographic Fisheries Investigations (CalCOFI), which is now more than forty years old.

Related, ecologically similar species—even ecologically similar complexes of species—are important in other eastern boundary currents, and analogous studies have been conducted in many parts of the world (e.g., Payne et al. 1987). My summary of research on the California stocks slights the contributions made by such studies to overall understanding, in the interests of brevity.

Much of the work I shall review has been based implicitly or explicitly on the assumption (called Hjort's first hypothesis) that the supply of food for the youngest larvae is both variable and crucial. Since these larvae (especially those of anchovy and sardine) have limited predatory capabilities, and since their rate of mortality is very high, it is a reasonable initial premise that variation in this mortality, caused proximally by variations in the food web and ultimately by variations in the ocean's climate, is an important source of variation in recruitment to the adult populations.

This premise stimulated great improvement in the capability to rear and study larval fish in the laboratory, including the cultivation of their food. To a degree, the premise has been statistically examined relative to other hypotheses, but the fact that it became technologically more feasible to study larval feeding than, say, predation on juveniles has lent the premise a certain pragmatic momentum. We often tend to continue longer than necessary the studies we already have developed the techniques to do, because development of new techniques to test other hypotheses, including pilot studies in the field, is frequently time-consuming and unrewarding.

First-feeding has been emphasized also because the need for specific, critical nursery habitats for young fish has been less precisely defined, and therefore is less likely to be investigated, for pelagic species than for demersal ones. Recent emphasis on the importance of hydrographic fronts, either as regions where larval food is produced or as regions where the circulation promotes cohesion of larval stocks (Sinclair 1988), is changing this.

I have chosen to review these studies (rather than, say, purely demographic studies of the fish populations) for the same reason: because linkages between the food web and recruitment have been explored. Only some of the explanations for variability in recruitment of these pelagic species explicitly invoke the food web, and in evaluating those that do, it is important to bear in mind some caveats.

First, the simplest depiction of a food web—as a chain from generalized phytoplankton to generalized zooplankton to young fish—is too simple. It is attractive (and therefore widely used) because the properties involved (total primary production, biomass of phytoplankton as chlorophyll, and zooplankton as displacement volume) are relatively easy to measure. However, phytoplankters and zooplankters come in various shapes and sizes, only some of which are actually eaten by larvae of a particular species (though pelagic marine animals are, in general, quite catholic in diet relative to terrestrial ones). For example, the CalCOFI program (which is one of the oldest programs in fisheries oceanography in the world) does not include routine measurement of the distribution or production of the specific food of any larval fish, though the types of plankton actually eaten are well known. Most correlative explanations of variability in larval feeding success rely on the assumption that the edible plankton is distributed more or less as are the more easily measured properties such as total chlorophyll or volume of large-bodied zooplankters.

Second, the distribution of sizes of plankters may well be influenced by the physics and chemistry of the environment, and this distribution has implications for structure of the web. One example is the competition of other zooplanktivores, particularly ctenophores or jellyfish, with fish for the same

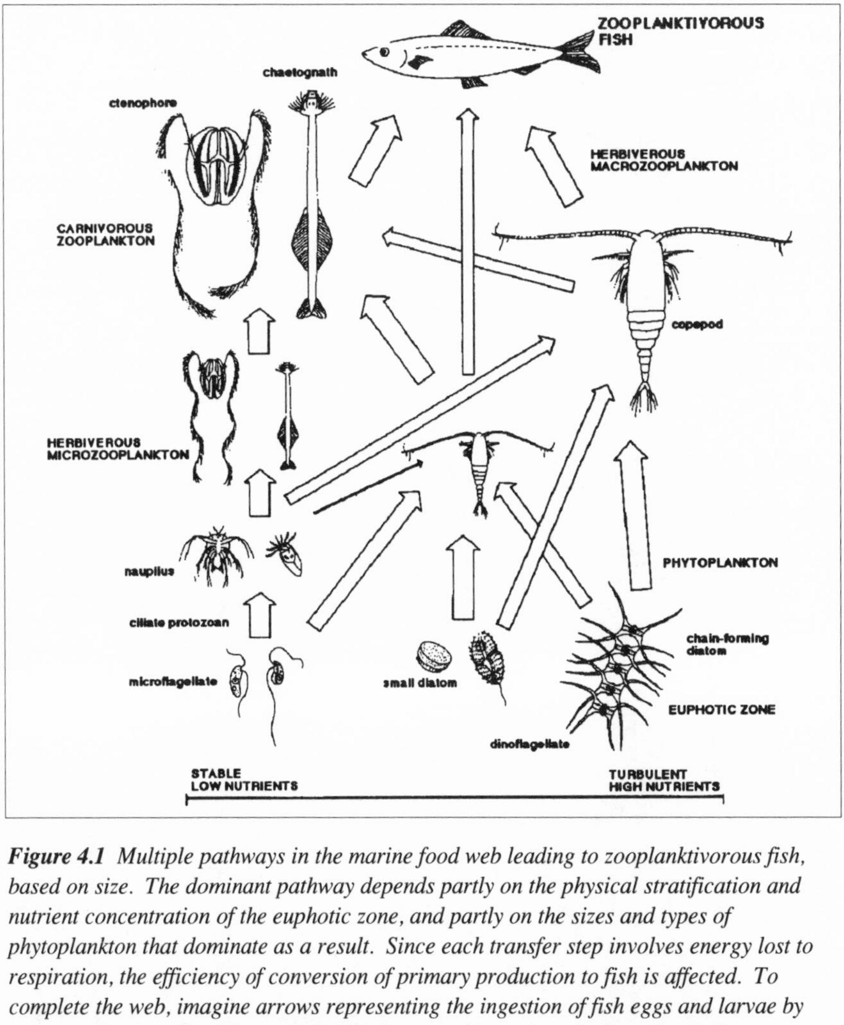

Figure 4.1 *Multiple pathways in the marine food web leading to zooplanktivorous fish, based on size. The dominant pathway depends partly on the physical stratification and nutrient concentration of the euphotic zone, and partly on the sizes and types of phytoplankton that dominate as a result. Since each transfer step involves energy lost to respiration, the efficiency of conversion of primary production to fish is affected. To complete the web, imagine arrows representing the ingestion of fish eggs and larvae by some copepods and carnivorous zooplankters, and ingestion of copepod eggs by some dinoflagellates. From Landry 1977.*

sizes of food (and the fact that many of these "competitors" are also larval fish predators), as suggested by Figure 4.1. A second example, due to improvements in techniques for assessing the abundances and rates of growth of bacteria, very small phytoplankters, and protozoans, is what has come to be called the "microbial loop," by which dissolved organic matter and detritus can re-enter the food web in which young fish participate (Figure 4.2). Cushing (1989) has reviewed the implications of the microbial loop for fisheries.

These concepts are less prominent in the work reviewed below than I

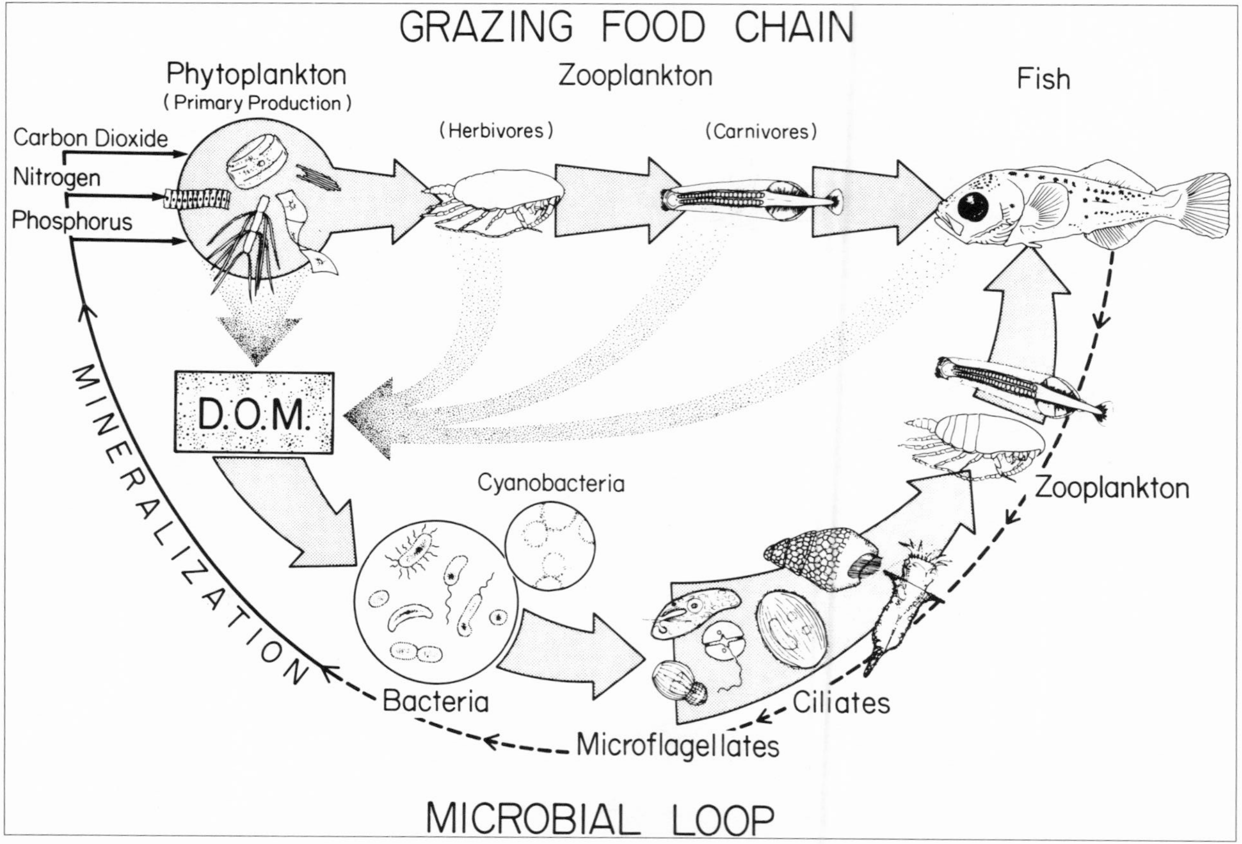

Figure 4.2 *Diagrammatic view, not necessarily to scale, of the classical grazing food chain and the "microbial loop," both ending in zooplanktivorous fish. The dissolved organic matter (D.O.M.) pool also receives input from the microbial loop, and elements of the grazing food chain also mineralize nutrients. Detritus (dead, particulate organic matter) is not shown. From Azam, unpublished, via Beers 1986.*

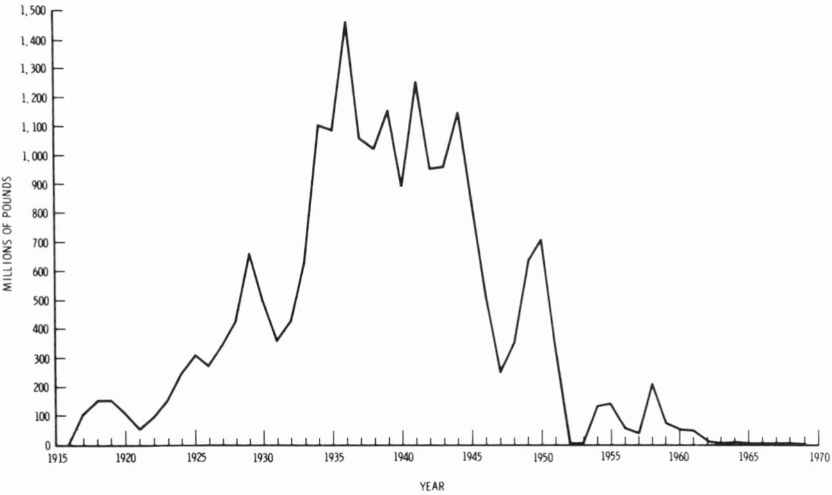

Figure 4.3 *Catch of sardines along the North American Pacific coast. From Frey 1971.*

believe they should be. To complete this picture, one might include the microcosms represented by organic aggregates, though (as discussed in Chapter 2) the significance of these to fish is virtually unknown.

Interannual to Interdecadal Scales

During the Depression, the California fishery for sardine began to grow rapidly until, by the mid-1940s, it had become, in tonnage, the largest fishery of its time. Abruptly, however, as the Second World War drew to a close, the sardine catch along the North Pacific coast declined dramatically. Within the next decade, it all but disappeared (Figure 4.3).

An intense argument developed, to a degree pitting federal against state fisheries scientists, as to the relative roles of overfishing and ecological change (the scientific argument was exacerbated by interagency rivalry; Radovich 1982, Scheiber 1990). Initial sampling indicated an increase in anchovy roughly matching the decline of sardine, suggesting a competitive replacement (Figure 4.4).

Scales then entered the story in two senses. Off Southern California there are rather isolated, deep basins, some of which are anoxic for long periods, meaning that the large benthic organisms that disturb the bottom sediments are rare or absent (but see Lange et al. 1987), and consequently fish scales and other sinking particles accumulate in distinct layers. By counting fish scales in cores of sediment, Soutar and Isaacs (1974) could study the

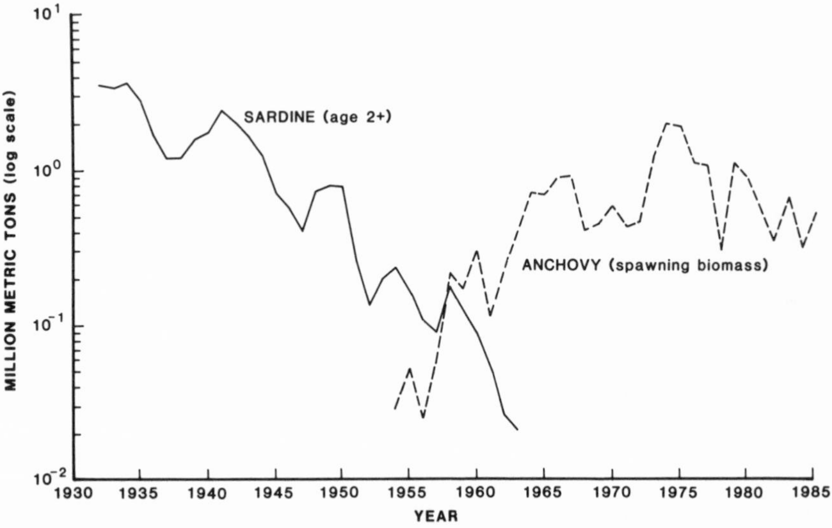

***Figure 4.4** Estimated spawning biomass (log scale) of sardine* (solid) *and anchovy* (dashed) *off California and northern Baja California. From various sources, summarized in McCall 1986.*

interaction between the species, at least the interaction indicated by their relative abundances, on a temporal scale that extended well before the period sampled by the fishery.

They found that the size of the sardine population (at least over the basin sampled) had been quite variable long before the intense fishery (Figure 4.5), and that the sizes of the populations were not in a simple, inverse relation to each other, as would have resulted from a purely competitive situation or predation by one species on another. If anything, the correlations between the scale deposition rates of the species were positive. Further, the total biomass of important pelagic fish of all types apparently had varied considerably on the scale of decades (Figure 4.6). This suggests again that changes in abundances of species were not due simply to variation in the partitioning of a constant, limited amount of food between the competing species of fish.

Several explanations for this variation are possible: that the relation between deposition of scales and population sizes had varied through time, so that the summed demand for food had been more constant than suggested by Figure 4.6; that some other complex of competing species, such as ctenophores or myctophids, also varied, inversely from the fish; or that the sizes of these populations had not been limited by food supply. This last possibility, though, is unlikely, since at its peak the sardine population alone was estimated to consume much of the secondary production of zooplankton in the

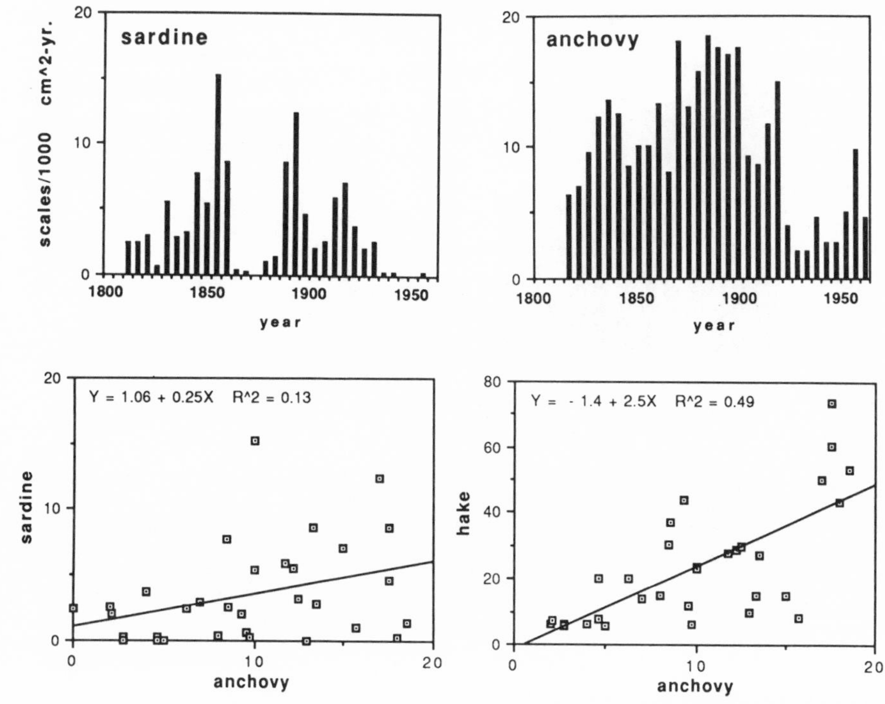

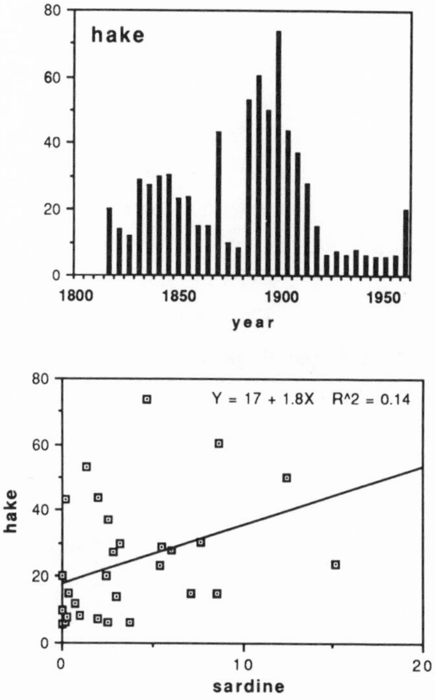

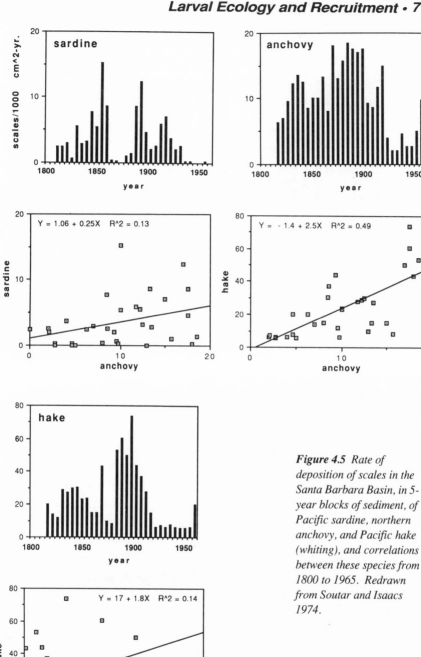

Figure 4.5 *Rate of deposition of scales in the Santa Barbara Basin, in 5-year blocks of sediment, of Pacific sardine, northern anchovy, and Pacific hake (whiting), and correlations between these species from 1800 to 1965. Redrawn from Soutar and Isaacs 1974.*

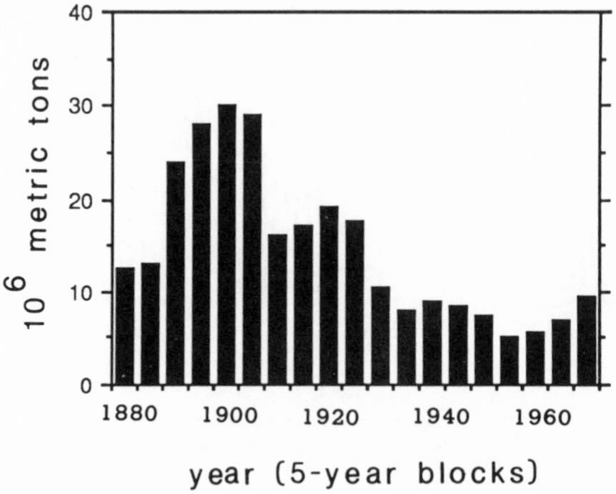

Figure 4.6 Biomass of the California Current coastal pelagic fish complex—sardine, anchovy, mackerels, saury, and hake—as inferred from current stocks and scale deposition rates (shown in Figure 4.5). From Smith 1978, via Bakun 1986.

California Current (Lasker 1970). Therefore, there may have been real variations on the decadal scale in the supply of food for these pelagic fish

The same sediments contain other informative materials, such as diatom frustules which yield information on interannual to decadal variation in pelagic organisms with much shorter generation times than fish and therefore with potential for more rapid variation (e.g., Lange et al. 1990). Direct comparison of variability of fish and their food has yet to be made from these records. Resolution of the finer scales is limited by the flux of identifiable remains and by the precision and accuracy of dating layers.

Variation in the population of anchovy, as revealed from scale records, is not related to the primary production of phytoplankton in the Southern California Bight (Figure 4.7) as estimated from temperature anomalies (which themselves may indicate variation in the supply of nutrients for phytoplankton). However, the biomasses of presumed anchovy food (macrozooplankton and microzooplankton) are statistically related to the primary production over shorter periods (Smith and Eppley 1982).

Interannual and interdecadal variation in pelagic fish stocks is of obvious concern to fishermen and fishery managers, but it impacts other other elements of the marine food web as well. Piscivorous seabirds should be particularly susceptible to such variation, as suggested by Figure 4.8 (and Figure 4.18), since their ability to switch to alternative food sources is limited by their diving capabilities. As questioned in Chapter 1, what are the causal connections (if any) between variability of fish stocks on these decadal scales (and, though this is an assumption, over large spatial scales as well) and biological-physical interactions via the food web, where shorter scales of variation are likely because of shorter generation times? What light can simulation modeling shed?

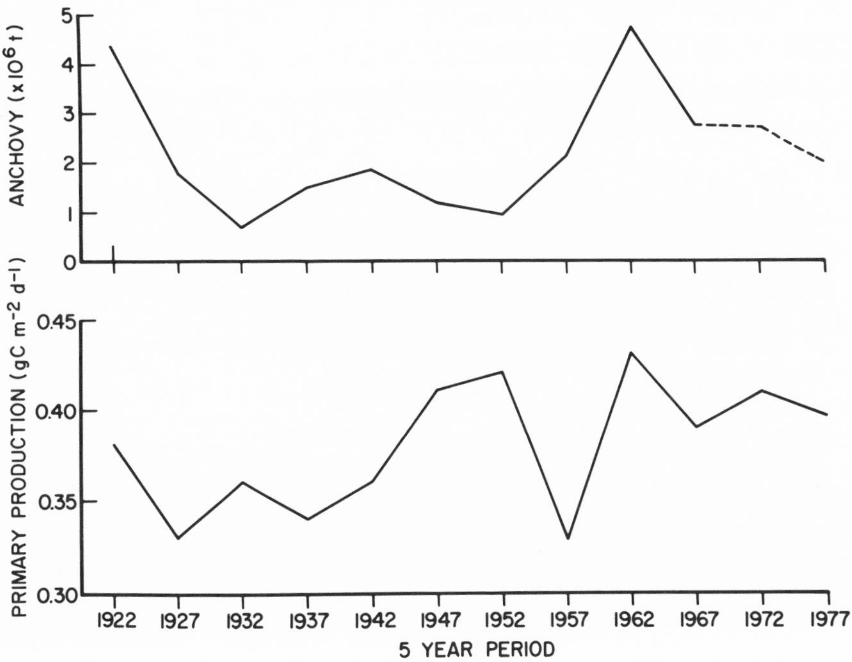

Figure 4.7 *Anchovy biomass (from scale deposition rate shown in Figure 4.5) and primary production of phytoplankton (from temperature anomaly and day length) in the Southern California Bight, in 5-year blocks, 1922–1979. From Smith and Eppley 1982.*

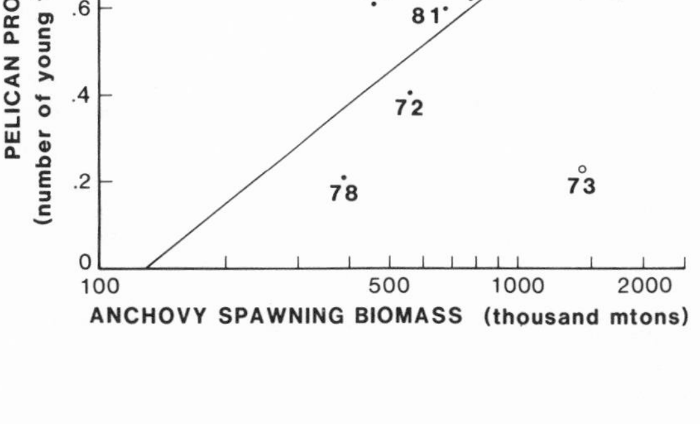

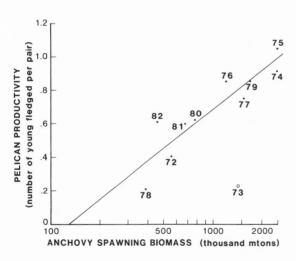

Figure 4.8 *Brown pelican productivity of young vs. spawning biomass of northern anchovy. From McCall 1986.*

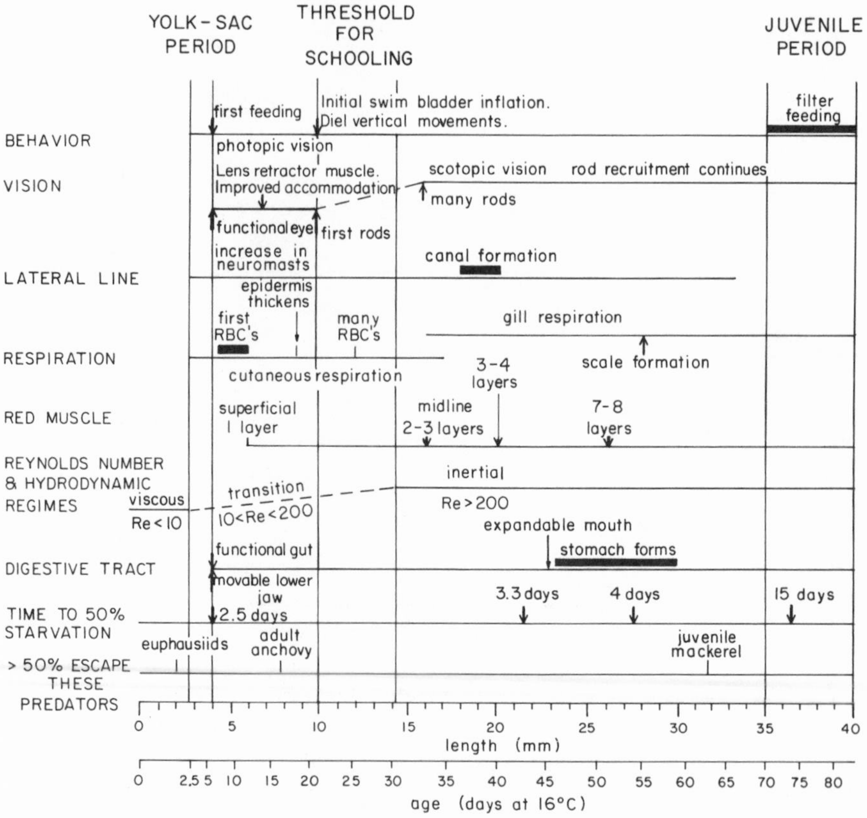

Figure 4.9 *Ontogenetic and behavioral events in northern anchovy larvae. RBC = red blood cells; time to 50% starvation = days of starvation at which 50% of the fish died. From Hunter and Coyne 1982. Bottom line shows the predators from which anchovy larvae are able to escape at least 50% of the time (information from Folkvord and Hunter 1986).*

The Smallest Scales

At the other end of the spatial/temporal scale is the range of seconds to days and millimeters to meters. This describes the range between the reaction of an individual larva to prey and the ambit of an individual during the few days after the yolk sac is absorbed and food must be found to avoid starvation.

Because it is possible to rear several pelagic fish in the laboratory and to determine the age of field-caught larvae from daily growth rings in their otoliths, it is also possible to observe and describe in detail the ontogenetic and behavioral changes in early life (Figure 4.9) and to focus on events at the scale of individual larvae. The ability to perceive and strike successfully at prey increases with age, as does the range of prey sizes that the larva can swallow

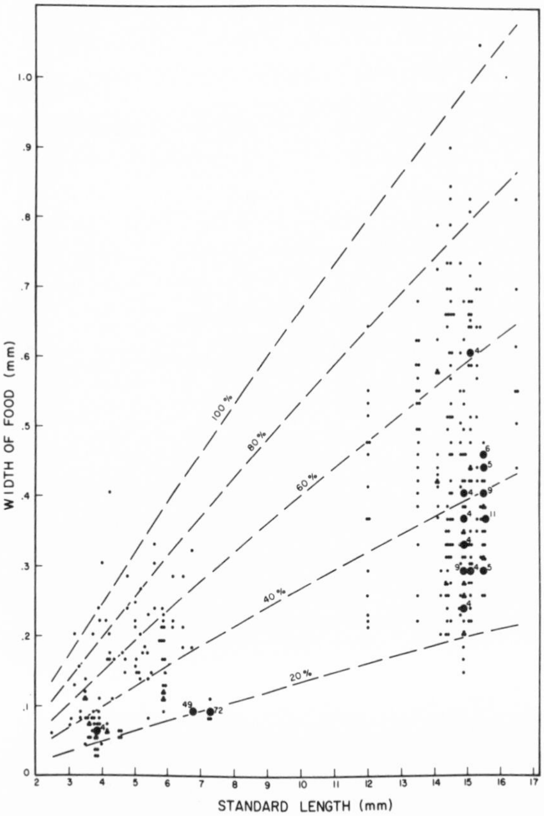

Figure 4.10 *Size of prey ingested vs. size of mackerel larva:* dashed lines, *widths of prey equal to percentage of mouth width. From Hunter and Kimbrell 1980.*

(Figure 4.10). Since eggs are spawned by schools of adults, the youngest larvae are quite patchy in distribution; they become more evenly distributed by dispersion (as well as rarer, because of mortality) until they begin to show their own schooling behavior (Figure 4.11).

Some experiments and observations on small scales have been directed at determining the factors affecting larval mortality due to predation by invertebrates and by other fish (including cannibalism), as shown in Figure 4.9. The ontogeny of larval escape responses presumably influences the age-specificity of larval mortality by interplay with the distributions of abundances of types of predators, much as prey distributions and the development of predatory capabilities by the larva affect the age-specificity of larval growth. The lighting conditions, presence or absence of alternative prey for the predator, and size (age) and health of the larvae all influence the susceptibility to predation (e.g., Lillelund and Lasker 1971, Folkvord and Hunter 1986, Butler and Pickett 1988; see Bailey and Houde 1989, for a review).

There has been little application of these laboratory studies in attempting

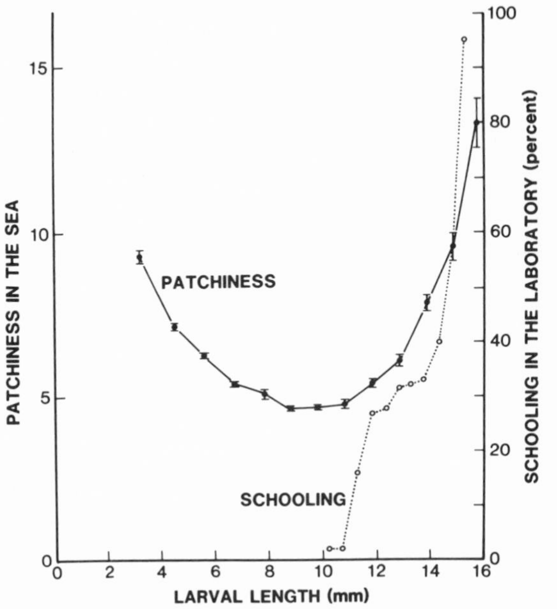

Figure 4.11 Patchiness of larval anchovy in the sea, as measured by Lloyd's index (the degree of departure from random dispersion) and the onset of schooling as measured by visual observation in the laboratory. See Figure 4.9 for conversion of length to age. From Hunter and Coyne 1982.

to account quantitatively for natural rates of mortality, though the large-scale distributions of potential invertebrate predators have been determined at certain times (Alvariño 1980). Pepin (1987) obtained some evidence indicating that larval *per capita* mortality was influenced by the relative availability to the predator of alternative prey (measured as zooplanktivorous fish biomass/zooplankton biomass).

The twin processes of becoming more efficient at predation and at avoiding predators interact to the degree that the growth rate affects the rapidity with which the larva attains the capability for strong escape (e.g., size-dependent mortality in the simulation by Wroblewski and Richman 1987, described below). What has not yet been accomplished is to integrate this type of information into a more general understanding of larval mortality in the ocean through knowledge of the distributions of larvae, their foods, and their predators, measured both on the scales at which the individual interactions occur and over sufficient space and time to pertain to the strength of recruitment of a year-class.

Well-designed laboratory experiments include reasonable duplication of natural conditions, limited by physical space and the investigator's motivation to simplify those properties thought to be extraneous and to exaggerate those of particular interest. Simulation models, though they also represent idealizations of nature by emphasizing certain properties, still permit extension in complexity beyond the laboratory results on which they depend, because they allow mathematical descriptions of individual components of behavior and reactions to specific environmental factors to be combined algebraically.

For example, the results of many separate behavioral studies of larval anchovy were combined with a description of abundances and distributions of their prey by Vlymen (1977) to assess conditions for larval growth. As noted in Chapter 2, laboratory rearing had suggested that larvae might require patches of dense prey for survival, rather than mean concentrations. Vlymen assumed that the cumulative abundance of prey of different sizes was hyperbolic (Figure 4.12)—large prey are rarer than small—and that horizontal patches of dense prey (as distinct from the layers discussed below) were superimposed on a background of dispersed prey, with the dimensions of the patches and the spacing between them depending on the "mean crowding" or contagion parameter from the negative binomial describing the patches and varying with size of prey (Figure 4.13).

Given the experimentally derived information on the searching behavior, attack success, digestion, and respiration of larvae of various sizes, Vlymen could then calculate the expected growth rates (excess of digestion over respiration) of larvae as a function of the patchiness of prey. As shown in Figure 4.14, growth is best at intermediate degrees of patchiness, where the density of prey of each size is approximately 10 times its "background" density. Lasker and Zweifel (1978) extended this model to explore further the consequences of the increasing success of attack and size of prey that can be ingested as a larva grows.

As is often the case, the result of Vlymen's model is not a prediction of larval growth useful to a fishery manager concerned about each year's recruitment, nor does it explain the large-scale variation shown in Figure 4.5. It does demonstrate, however, that information on the scales of patchiness of prey of various sizes, as well as total abundances, may be necessary (but not sufficient) for such a prediction or explanation. Indeed, the work of Owen (1989), reviewed in Chapter 2, is a good example. Specifically, Owen found that the intensity of patchiness of about 20% of the cases analyzed was sufficient to maximize larval growth by the criterion shown in Figure 4.14.

The utility of this model for a manager, and understanding of the larval environment, would be enhanced by a description of the relation between the contagion parameter, k, and physical turbulence generated by weather acting on a water column whose vertical stratification (and therefore the depth of penetration of turbulence) is also predictable from routine measurements. Simulations such as that by Davis et al. (1991), which I reviewed in Chapter 2,

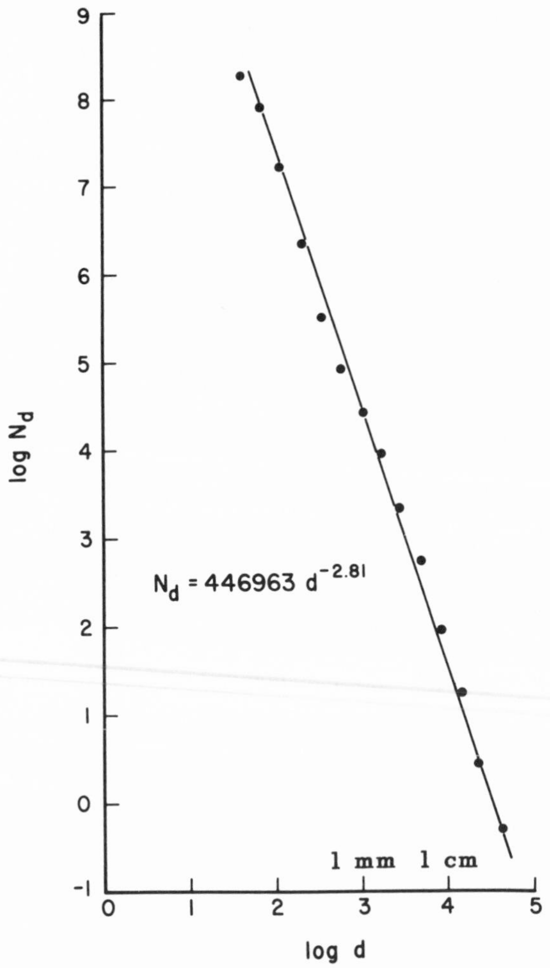

Figure 4.12 Cumulative distribution of sizes of particles, assumed to represent potential food within the range of sizes larval anchovy can ingest, where N_d is the concentration of particles (ml^{-1}) greater than d µm in diameter. From Vlymen 1977.

are aimed at establishing analogous relations. Do stratification and turbulence vary on scales that affect long-term changes in pelagic stocks? If they do, does this variation really act through patchiness, through changing the primary production at the base of the food web, or through some other mechanism? For the field scientist, it becomes interesting to know whether the spawning of the adults is at all influenced by the patchiness of larval prey, either as sensed directly by the adults or as regulated by topographic features that the adults perceive. Does the growth rate of larvae at various ages, as measured from otolith rings, vary with patchiness of prey? This is a different question from asking whether mean growth rates and mean concentrations of prey are correlated at large scales in space or time.

I should point out here that weather, and water movement in response to

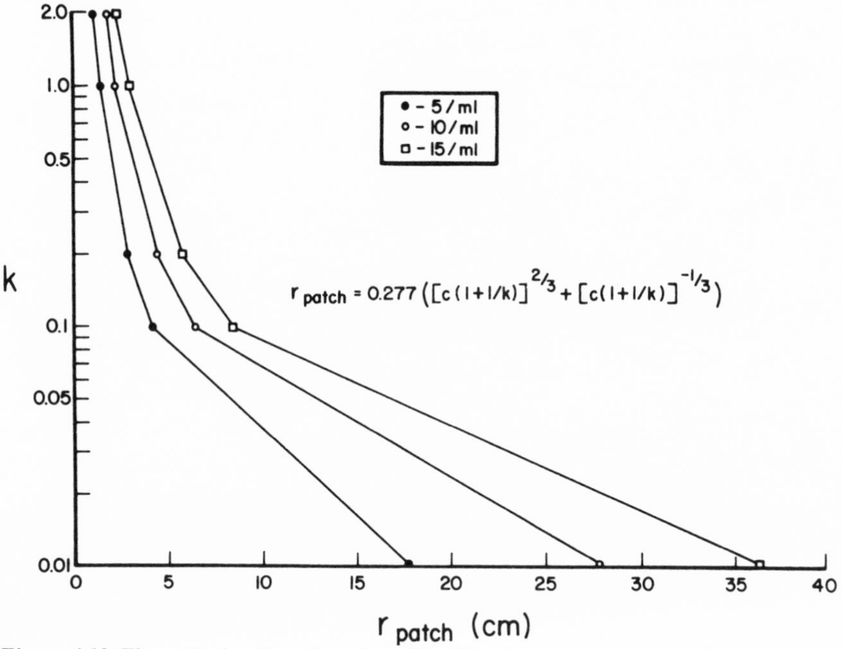

Figure 4.13 *Theoretical radius of patches of food for three concentrations of food particles, as a function of the contagion parameter, k, from the negative binomial distribution. From Vlymen 1977.*

weather, may have effects on larval fish that are less directly due to the food web and its distribution on small scales. One hypothesis states that advection of larvae from favorable to unfavorable areas by wind-generated flow can vary interannually ("favorable" and "unfavorable" are imprecisely defined, but are large in scale). Indeed, it seems to me that most studies of the effect of physical oceanographic processes on pelagic fish recruitment, when the physical oceanography is explicit—not simply implied, as in Cushing's (1972) match/mismatch hypothesis—have focused on the advection of larvae.

For a coastal species, advection offshore by upwelling is usually thought to be unfavorable, even if the quality of "badness" cannot be defined. Seasons and locations of spawning are thought to have been influenced by evolution to minimize such drift (Parrish et al. 1981, Power 1986). Yet, a map of maximal abundances of nauplii in the California Current (Figure 4.15) suggests that offshore advection could bring nauplivorous larvae into better rather than worse conditions, especially north of Point Conception. (This impression results in part from the paucity of very nearshore samples in the years mapped.)

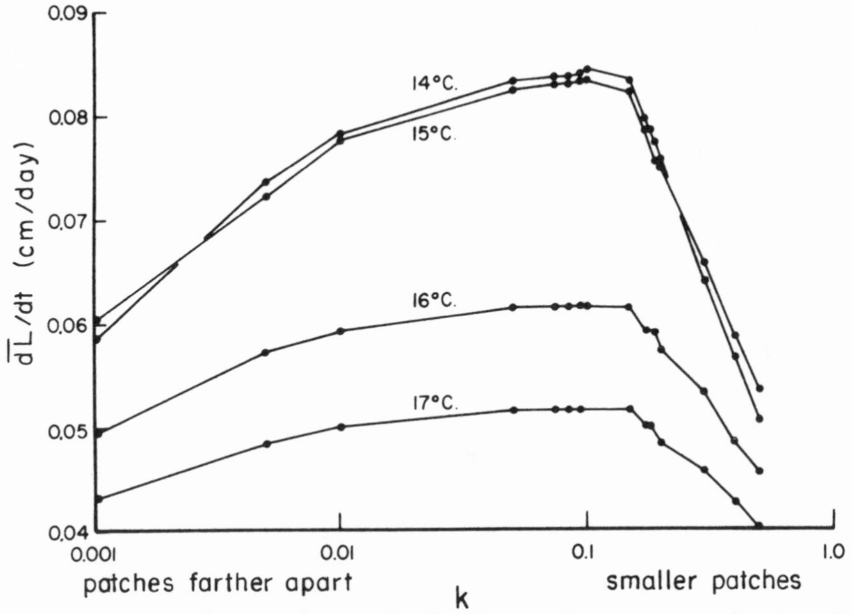

Figure 4.14 *Growth rates in length (cm day $^{-1}$) of larvae at 4 temperatures as functions of the prey contagion parameter, k. See Figure 4.13 for relation of k to patch radius. From the relation between prey concentration and size (Figure 4.12), a given k may represent different patch radii and different spacing between patches for different sizes of prey. From Vlymen 1977.*

In the discussion that follows, the reader may recall the complaint of experimentalists, summarized in Chapter 1, that even strong correlations in field data may be ambiguous as to causation unless the various plausible causes act on different scales and the correlations are scale dependent.

Vertical Layering and the Event Scale

The issue of small-scale distributions of food, emphasizing vertical layers rather than horizontal patches, has been explored because of laboratory evidence that certain large dinoflagellates are suitable food for first-feeding larval anchovy (the more robust larvae of other species can feed on microzooplankton), and may be the only suitable food occurring naturally in subsurface layers dense enough for their growth. Lasker (1975), sampling water and seston from various depths off southern California, assayed the samples for suitability by determining in which samples first-feeding larval anchovy (hatched on shipboard) could fill their guts with food. Only in a subsurface chlorophyll maximum layer (which contained large dinoflagel-

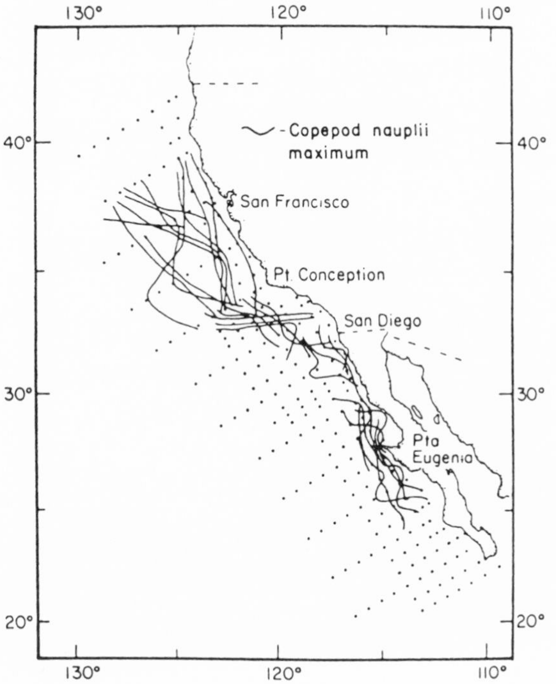

Figure 4.15 *Positions of maxima in naupliar copepod abundances for individual cruises, June 1949 to July 1951. From Arthur 1977.*

lates) was this possible. Kiefer and Lasker found evidence of this layer (or similar ones) at six stations for 100 km along the 20 m contour of the Southern California Bight.

A sudden windstorm, usually the bane of the seagoing oceanographer, apparently disrupted this layer (Figure 4.16), and at no depth was there successful feeding. Although advection of a new water mass with new, more homogeneous vertical distributions was not precluded, the result was consistent with the hypothesis that mixing by storms, often thought to be beneficial in restoring nutrients for phytoplankton to surface layers, is actually detrimental to larval anchovy because it destroys essential dense layers of dinoflagellates, even if the average biomass of phytoplankton increases. As noted in Chapter 2, the smallest-scale turbulence associated with the storm could also have been directly detrimental to the continued growth of the dinoflagellates, but this would be unlikely to affect the larvae for some days if the dense layering had persisted.

Two scales are involved here (only two, since first-feeding larval anchovy lack scales - Figure 4.9): the temporal "event" scale of the storm and the vertical scale of a few tens of meters. Of course, coastal storms vary in intensity and effect. A similar study of vertical distributions by Mullin et al. (1985), also interrupted by the mixed blessing of a storm, revealed no such

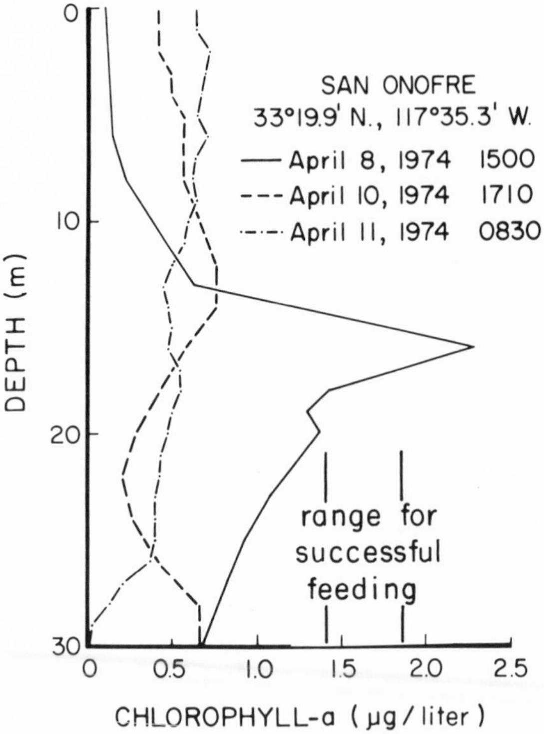

Figure 4.16 *Vertical profiles of chlorophyll off southern California before a windstorm (8 April) and after it (April 10 and 11). Vertical bars indicate the range of chlorophyll concentrations where shipboard experiments indicated successful feeding by larval anchovy on large dinoflagellates. From Lasker 1975.*

layering of food, either for larval anchovy or for mackerellike larvae feeding on larger prey (Figure 4.17).

It is not true that all layers of large dinoflagellate cells are beneficial to larval anchovy. Lasker (1978) noted that the armored dinoflagellate, *Gonyaulax polyedra,* is less satisfactory as a food source than the unarmored *Gymnodinium splendens*. There is even evidence from studies of the feeding and distribution of zooplankton that some *G. splendens* layers may be poor feeding locations (Fiedler 1982). Around the North Sea, subsurface layers may be dominated by species of dinoflagellates known to be toxic (e.g., Bjornsen and Nielsen 1991, discussed in Chapter 2). Thus, there is some asymmetry in predictiveness. Although conditions that prevent the formation of dense layers of dinoflagellates might thereby prevent successful recruitment, conditions that permit such layers might not ensure strong recruitment unless the particular species responding to the conditions was of the right size and nutritional quality for the larvae.

These small scales are related to the interannual scale of variable recruitment through interannual variation in weather. Supporting evidence that years of intense mixing by local winds might be poor for recruitment of

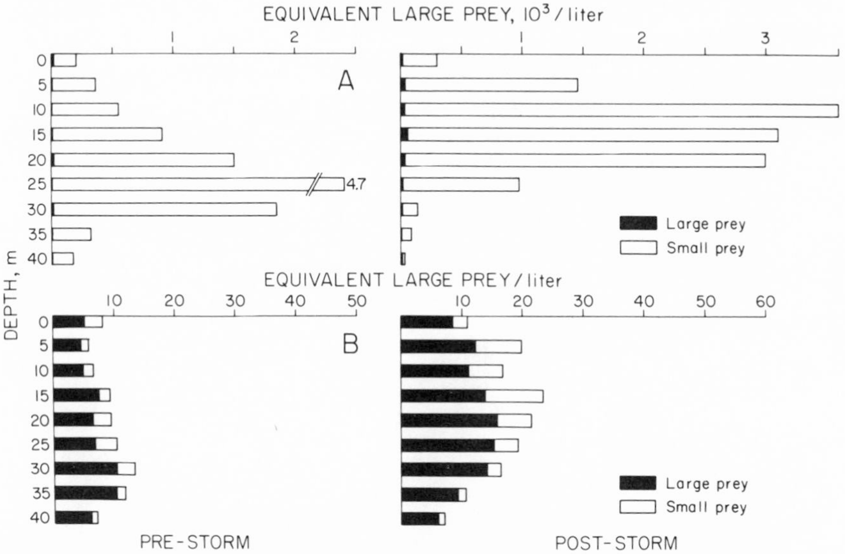

Figure 4.17 *Vertical distributions of prey off southern California before and after a windstorm, for* (A) *anchovylike and* (B) *mackerellike larvae. Within each type, prey are categorized as large or small ("large" prey for anchovy is approximately "small" prey for mackerel), and concentrations are expressed as equivalent large prey. From Mullin et al. 1985.*

anchovy was obtained by examining the reproduction of the murrelet, a seabird much dependent on young-of-the-year anchovy (Lasker 1978). Coastal winds in December–February (anchovy spawning period) were strong in 1975–76, weaker in 1976–77, and strongest in 1977–78; March–April winds were much stronger in 1976 and 1977 than in 1978. The timing and extent of clutch production by the Xantus' murrelet (Figure 4.18) were consistent with the hypothesis that larval anchovy survival was much better in 1977 than in 1976 or 1978. However, direct interference by winds with the birds' nesting, suggested by the difference in timing between 1976 and 1978, could not be ruled out, nor could variations in spawning by adult anchovy or depth of occurrence of the young fish.

Further evidence showing that both winter storms and spring advection of larvae could be significant was obtained by comparing the birthdates of larval anchovy surviving to be at least six months old (based on daily growth rings on their otoliths) with the dates of spawning as determined by surveys (Methot 1983). In 1977–78, the winter was stormy but the spring relatively mild (see above), and large-scale winds of the appropriate direction to cause transport of larvae offshore were anomalously weak. The actual production of larvae by spawning from an adult biomass of 1.3×10^6 tons peaked in late

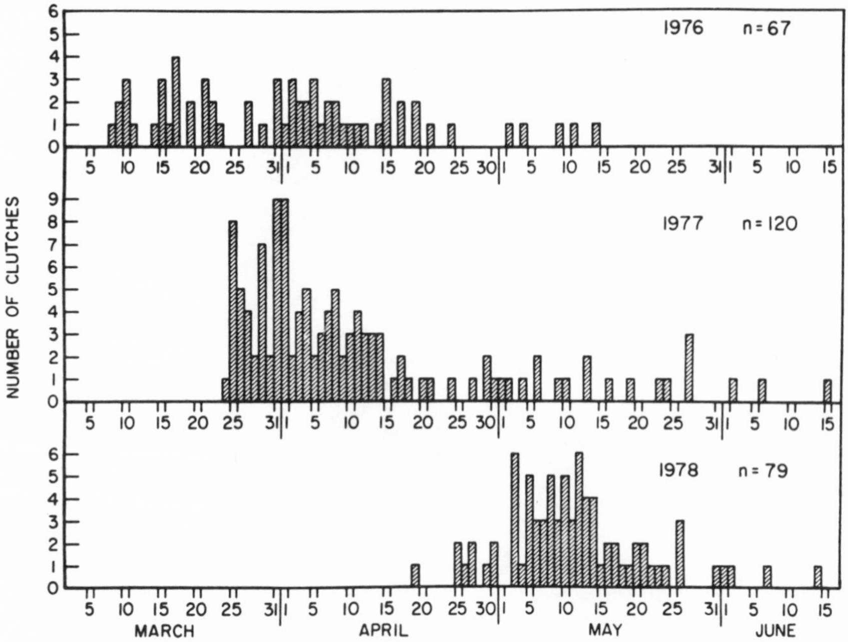

Figure 4.18 Clutch initiation dates in 1976, 1977, and 1978 for Xantus' murrelet on Channel Islands in the Southern California Bight. From Lasker 1981.

February, but the larvae that actually survived to recruit were born primarily in March and April (Figure 4.19), and the overall strength of the year-class was good. Hence, whatever harm the winter storms had done was apparently compensated by good survival of larvae born late in the season. By contrast, the birthdates of survivors more nearly matched the timing of spawning in 1978–79, when both the winter storminess and the spring winds causing offshore transport were normal, but the year-class strength from a spawning biomass of 1.7 x 10^6 tons was only half that of the previous year. One wonders what the year-class strength in a year of both unusual storminess and offshore transport would be.

Simulation models again have a dual role: they show directly, though in an idealized situation, the various effects of mixing by wind (nutrients mixed to the surface, layers dispersed); and they allow "experiments" to be performed by altering wind speed and/or duration or by changing the mathematical formulations used to express ecological relations. Wroblewski and Richman's (1987) model of the responses of a simple food chain of nutrients-phytoplankton-zooplankton-larval fish illustrates this, though the model is restricted to vertical distributions (i.e., no horizontal variation) and is devoid of migratory behavior by the organisms and sinking of particles. It assumes

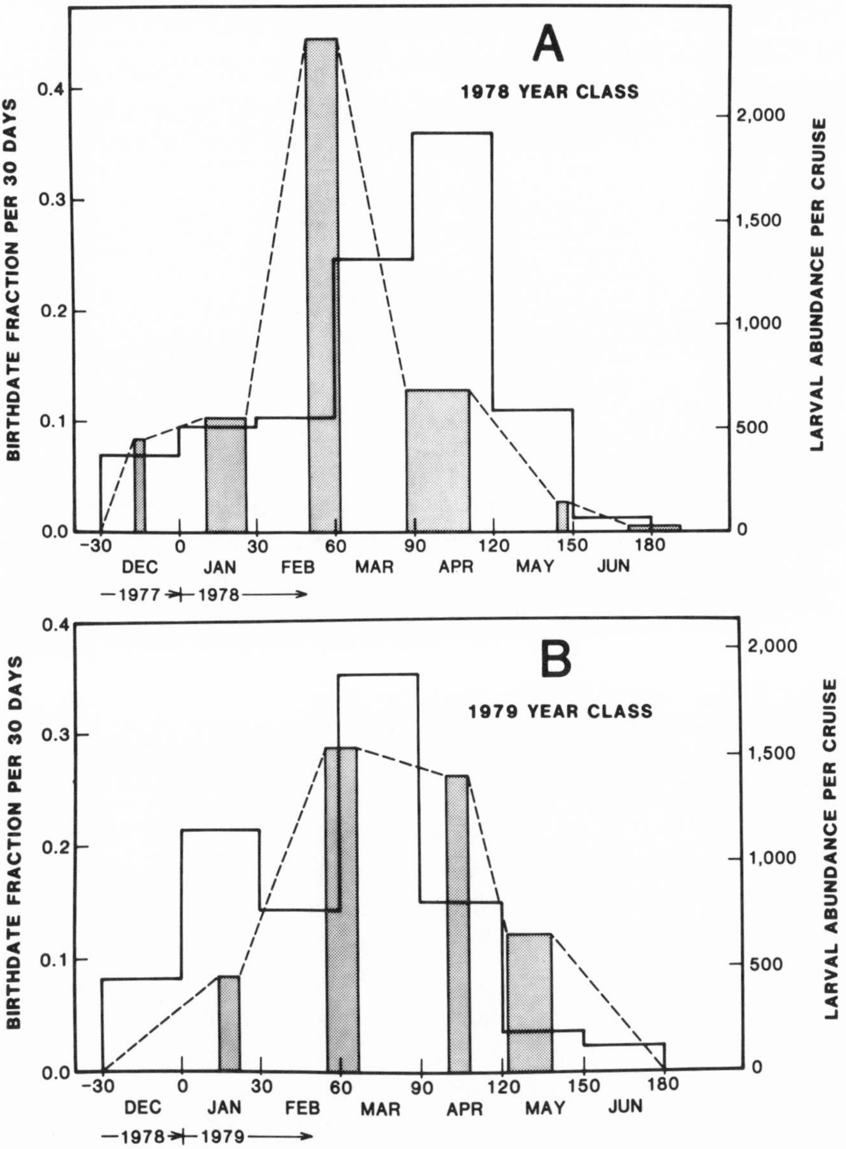

Figure 4.19 *Birth dates of juvenile anchovy surviving at least 6 months (open histogram, left axis), and production of larvae (shaded histogram connected by dashed line, right axis), in* (A) *1978, when spawning biomass was* 1.3×10^6 *tons, and* (B) *1979, when spawning biomass was* 1.7×10^6 *tons. From Methot 1983.*

that all phytoplanktonic biomass is equally good food for herbivores, and it lacks both the smaller-scale effect of turbulence on contact between individual larvae and particles of food (see Chapter 2) and a dynamic description of predators on larvae.

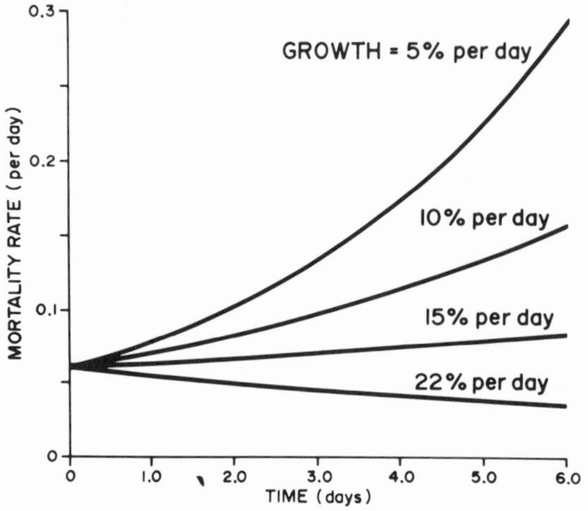

Figure 4.20 *Assumed mortality rates of larval anchovy growing at different rates. From Wroblewski and Richman 1987.*

Based on the qualitative argument that slowly growing larvae are more subject to predation than rapidly growing ones, the mortality rate was taken as a function of larval growth rate and age (Figure 4.20). Growth rate was calculated as assimilation of food (dependent, of course, on food availability) minus respiration (excretion). The phytoplankton grows according to availability of nutrients (which therefore decrease) and light, and is eaten by zooplankton, which itself grows in response but dies at a constant rate (i.e., predation by larvae is insufficient to affect the zooplanktonic crop; cf. McGowan and Miller 1980, Cushing 1983).

In the absence of wind, the vertical distributions change due to warming at the surface and depletion of nutrients by phytoplankton (Figure 4.21); a one-day windstorm restructures these distributions somewhat, and ecological interactions tend towards restratification (Figure 4.22). Larvae (which feed on a fixed fraction of the zooplanktonic crop, not on phytoplankton) are "spawned" into the system just before the storm, so that their mortality is uniform with depth and the vertical distribution of their growth and mortality rates changes because of the dynamics of the simple food chain (Figure 4.23).

Such a model is not intended for prediction of actual events; rather, it is a tool to explore ecological relations and reveal critical gaps in knowledge. For example, the model permits comparison of the effects on larval mortality of windstorms of various durations (differing, real events) with those of grazing on phytoplankton by zooplankton (different formulations of the same process). Wind has an effect on larval survival and growth, but the differences between a one-day and a three-day storm are minor compared with the issue of the correct description of the dynamics within the simplified food chain.

In a subsequent version, Wroblewski et al. (1989) demonstrated the

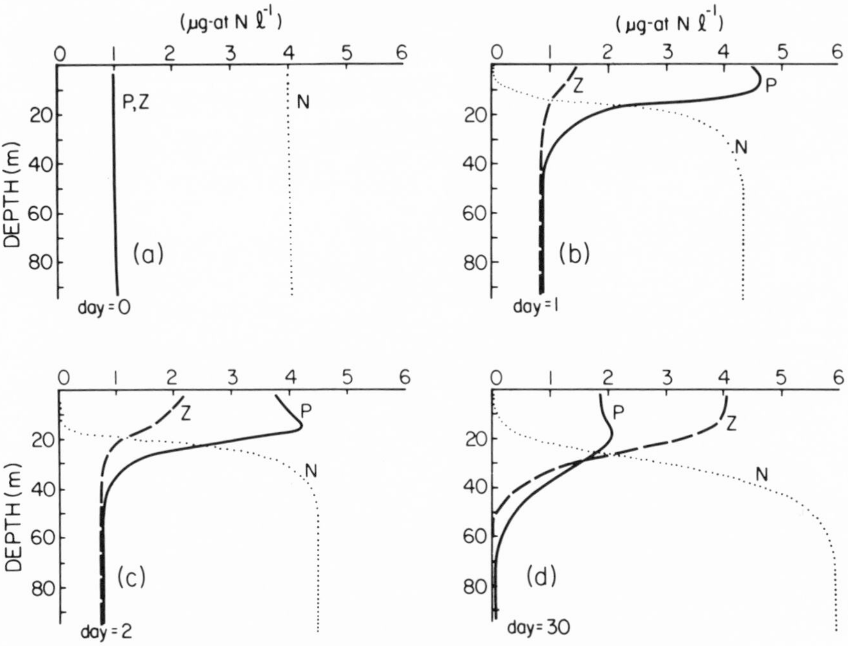

Figure 4.21 Evolution of vertical distributions of phytoplankton (P), zooplankton (Z), and nutrients (N), expressed as concentrations of nitrogen, for 30 days in the absence of wind. From Wroblewski and Richman 1987.

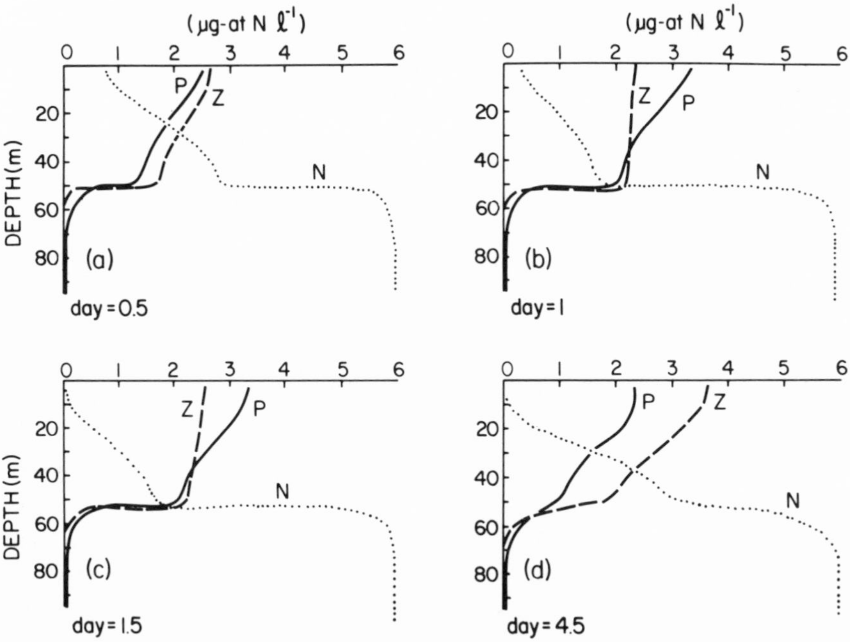

Figure 4.22 *Starting at the period of no wind shown in* d *of Fig. 4.21, evolution of vertical distributions* (a) *0.5 day,* (b) *1.0 day,* (c) *1.5 days, and* (d) *4.5 days after a wind of 10 m* sec^{-1} *for 24 hours. From Wroblewski and Richman 1987.*

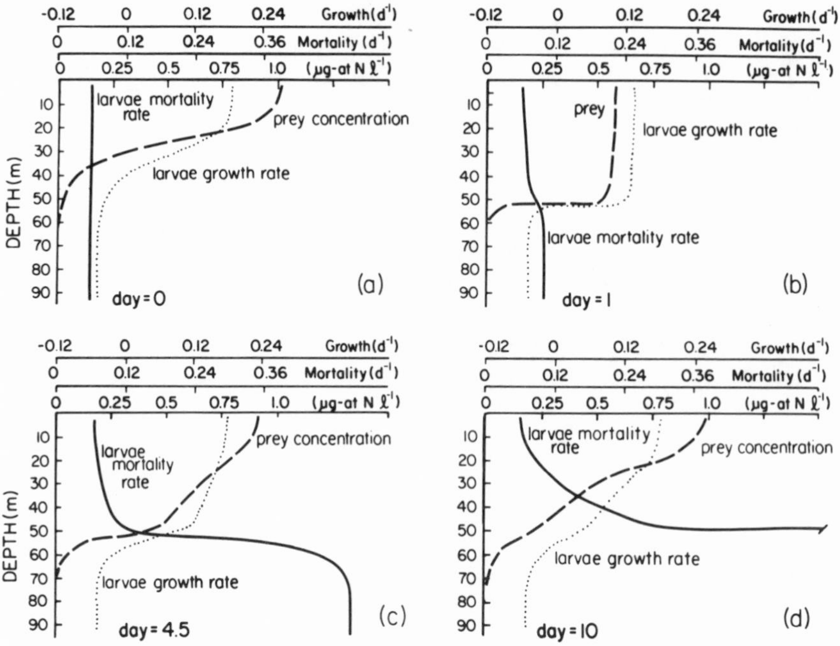

Figure 4.23 *Vertical distributions of prey concentration (proportional to Z in Figure 4.21), larval growth rate, and larval mortality rate (initially uniform; see day 0 in Figure 4.20)* (a) *before,* (b) *immediately after,* (c) *3.5 days after, and* (d) *9 days after a wind of 10 m sec*$^{-1}$ *for 24 hours. From Wroblewski and Richman 1987.*

intuitively reasonable proposition that a sequence of storms is quite deleterious, increasing larval mortality to 21% per day by preventing restratification of food. If concentrations of larval food are initially suboptimal, however, a single storm followed by at least two weeks of calm weather can be beneficial to those larvae born such that they begin feeding a few days after the storm, though it is disastrous for those larvae that begin to feed just before the storm strikes (Figure 4.24).

Relevance to the Recruitment Problem

Returning to the world of real data, recent investigation of the interannual variability in mortality of larval anchovy, and of recruitment to the adult stock, has helped place the issue of storms and layers of abundant food in larger perspective. It is worth re-emphasizing here, given my comments in Chapter 1, the enormous amount of data, on one species of fish alone, which must be gathered before competing hypotheses can be tested. If wind-induced mixing is somehow detrimental to young anchovy larvae (through disruption of

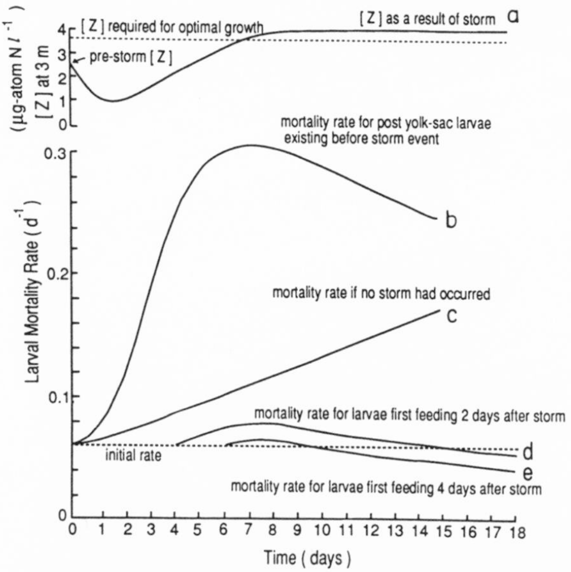

Figure 4.24 Mortality rates of larvae at 3 m depth for wind events of 20 m sec^{-1} *for 48 hours (days 0–2).* Top: a, *zooplanktonic biomass (as N vol*$^{-1}$*) as a result of the storm (stimulated by nutrients), compared with that required for optimal growth or larvae.* Bottom: b, *mortality rate for larvae that start feeding as the storm begins;* c, *larval mortality rate without storm;* d *and* e, *mortality rates for larvae which begin feeding 2 and 4 days after the storm has passed. From Wroblewski et al. 1989.*

layers of food, or some other mechanism), then larval mortality ought to be statistically better related to storminess than to other possible sources of mortality. To test this fully requires that the other sources of mortality also be independently measured, and such data are seldom available.

However, the importance of storms relative to other (though unknown) sources of mortality can still be examined. Peterman and Bradford (1987) calculated, for January to April for 13 years between 1954 and 1984, the number of four-day periods in each month for which wind speeds were less than 10 m sec^{-1} (minimal turbulence). They then weighted the value for each month by what fraction of total spawning for the year occurred in that month; that is, each month's "importance" depended on how much of the total year's spawning occurred then. For these 13 years, there were also estimates from samples of the daily mortality of 5–19-day-old larvae (though, of course, mortality probably varied both temporally and spatially within each year),

Figure 4.25 shows that average mortality decreased as the frequency of calm periods increased. This was only true when the frequency was weighted by each year's temporal distribution of spawning, implying that the really significant issue was the frequency of calm periods when spawning was most intense. The result does not prove, of course, that the mechanism of mortality involved layers of food; such proof would have required data on frequency of starvation (now obtainable from histological examination of specially preserved larvae) or on larval growth rate (from the daily rings of the otolith, also requiring special preservation). However, Peterman and Bradford did show

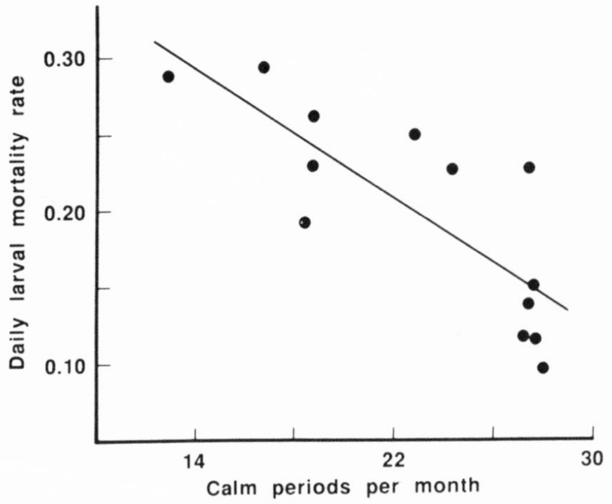

Figure 4.25 Average instantaneous mortality rate of northern anchovy larvae in each of 13 years, vs. summed number of calm periods (4-day stretches of winds <10 m sec^{-1}) per month in the winter–spring spawning season for each year, weighted by the fraction of each year's spawning which occurred in that month. From Peterman and Bradford 1987.

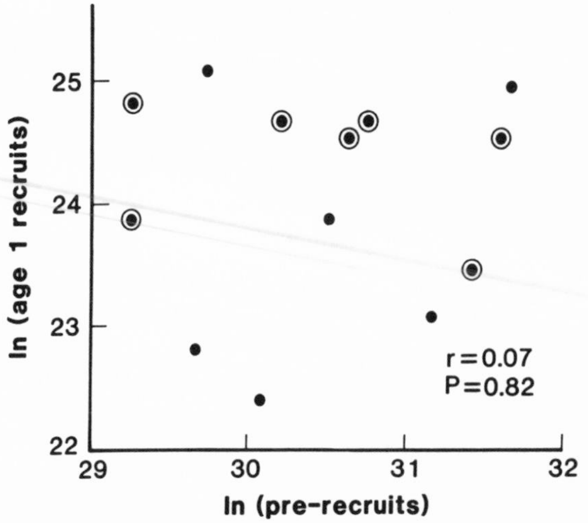

Figure 4.26 Abundances of 1-year-old anchovy recruits vs. abundances of 19-day-old prerecruits for 1965–1985 year-classes (both log scale). Relation is quite nonsignificant. From Peterman et al. 1988.

that larval mortality was related neither to the size of the adult anchovy population (thus, cannibalism was not dominant) nor to offshore transport driven by the larger-scale wind field.

This does not mean that predation and offshore transport are insignificant as variables for all ages of young anchovy, nor that interannual variation in recruitment is due primarily to larval mortality. In fact, the annual recruitment, which varied about tenfold from 1965 to 1985, bore little relation to the abundance of 19-day-old larvae or of eggs (Figures 4.26 and 4.27, Peterman et al. 1988). Note that the period studied includes the years for which Lasker (1978) argued that recruitment of anchovy, as reflected in seabird breeding (Figure 4.18), was related to the wind field, and concluded that variation in first-feeding survival affected recruitment.

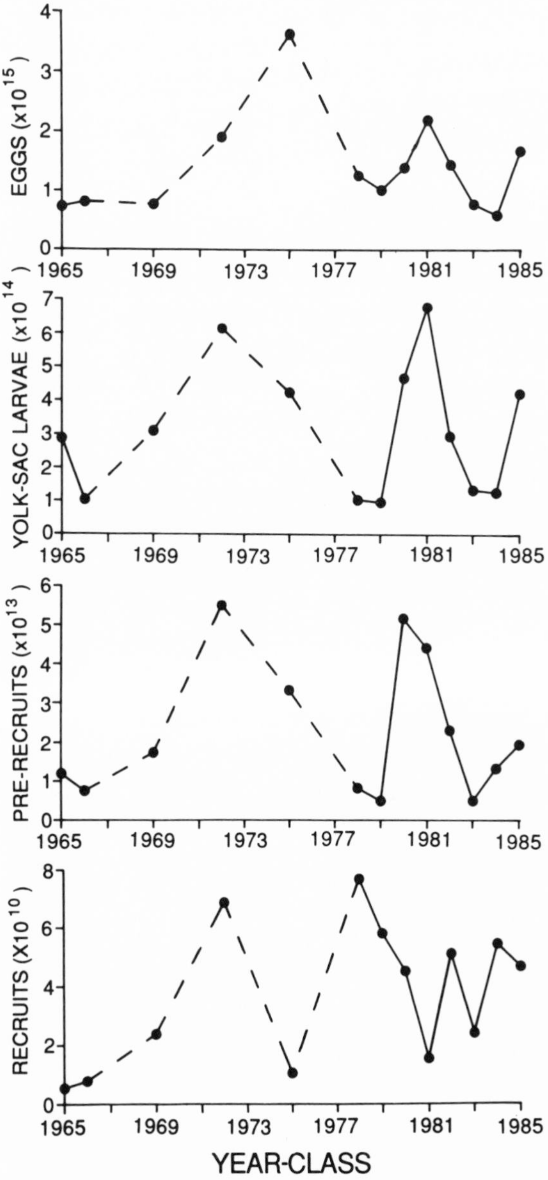

Figure 4.27 *Estimated abundances of early stages of northern anchovy, by year-class. Before 1978, abundances were estimated only every third year; broken lines connect these points but may not represent actual abundances in the intervening years. From Peterman et al. 1988.*

This is, clearly, a disappointing and anticlimactic conclusion. It implies at best that larval trophic relations, and the small-scale and event-scale processes that regulate them, have rather little bearing on the manager's problem of predicting recruitment, however interesting they may be in their own right. Of course, recruitment in other species may prove to be more responsive to early larval mortality, but given the limited capabilities of first-feeding anchovy relative to other larvae, any stronger relation between first-feeding success and recruitment would probably be due to relative constancy of mortality in the juvenile period for such species, rather than to more variable food-related mortality in larval stages.

Space on the Mesoscale

Physical processes and planktonic biomasses vary horizontally within the anchovy's spawning range and within one season on the mesoscale (a few to a few hundred kilometers). Larval growth and survival might be expected to vary on the mesoscale as well, but such variation has been very difficult to demonstrate. First, the time scale of the larval parameter being measured must be appropriate for the duration of residence of larvae in the mesoscale feature hypothesized to be causal. Second, the larval parameters have both analytical and real, smaller-scale variability associated with them, and an enormous amount of work can be required to demonstrate significant variability on the mesoscale.

Morphometric, histological, and biochemical measures of larval "condition" have proven useful. For example, Hakanson (1989a, b) recently showed, by rearing larval anchovy at various concentrations of food and starving them for various periods, that of the lipid components, triacylglycerol is particularly responsive to nutritional conditions, while cholesterol and polar lipids are stabler (as would be expected from their structural roles). He was able to demonstrate significant mesoscale variability within a spawning season and a correlation of variability to the lipids of large zooplankters (and, by inference, to growth and production of larval food), but he did not show a comprehensible spatial pattern or develop a statistical model for predicting good larval condition from other measurements.

A type of mesoscale feature that has received considerable attention along the U.S. east coast is the system of eddies generated by a meandering current such as the Gulf Stream (discussed in Chapter 3). The effects of similar features in the California Current on the distribution of zooplankton have been investigated (Haury et al. 1986), but less is known about the relations between such eddies and larval fish, either in the sense of advecting the larvae or through modifying their food web. However, Fiedler (1986) argued that, in 1985, a 100–200 km northeastward displacement of an anticyclonic eddy (normally centered at 32°N, 124°W), together with offshore expansion of the spawning range of adult anchovy, was responsible for anomalous offshore transport of larvae by entrainment of water from the

Southern California Bight into the eddy. The fate of these larvae was unknown, and overall recruitment of the 1985 year-class was not adversely affected.

Related studies have been conducted in the western North Atlantic. Flierl and Wroblewski (1985) showed through a simulation model how warm-core Gulf Stream rings could impose mortality by entraining larvae from the U.S. eastern continental shelf. Supporting evidence, from analysis of interannual and geographic variability in recruitment success of demersal fish stocks, indicates negative relations between recruitment and indices of offshore entrainment by the rings during the larval period (Myers and Drinkwater 1989). Interestingly, variability in recruitment of pelagic fish stocks was not related to the incidence of rings, but Myers and Drinkwater pointed out that the estimates of recruitment to these stocks are not very reliable.

Predictable mesoscale hydrographic regimes created by bottom topography, such as tidal fronts, are less common along the California coast than in regions occupied by Atlantic herring stocks (see Sinclair 1988, and Chapter 3) because the Californian shelf is so narrow and tidal currents relatively weak. Nevertheless, the offshore islands and banks that serve to separate the Southern California Bight from the California Current proper form a habitat on this spatial scale which seems to provide an enhanced supply of food for some larvae. Theilacker (1986) showed through histological analysis that the incidence of starving first-feeding jack mackerel larvae is much smaller around these islands than in the open California Current (Figure 4.28). Presumably (but not demonstrated in the same study), the supply of food is greater around the islands than farther offshore. An alternative, though more complex, hypothesis is that predators on larvae weakened by starvation are much more active around the islands, rapidly removing such larvae from the population there. Calculation of total larval mortality rate would resolve these alternatives.

Indeed, the contribution of starvation to total mortality of jack mackerel larvae in the offshore region, but not around the islands, was investigated by Hewitt et al. (1985), who used laboratory growth rates and daily growth rings of otoliths to estimate larval ages, and assumed that production of eggs was constant over the period represented by the range of ages of the larvae. Mean larval growth rate was fastest in the subarea where biomass of zooplankton was highest, but differences in growth rate were not statistically significant, and there was no apparent relation between zooplanktonic biomass and incidence of starvation in the larvae. Mortality rate was calculated as a function of larval age, though extensive manipulation of data concerning the

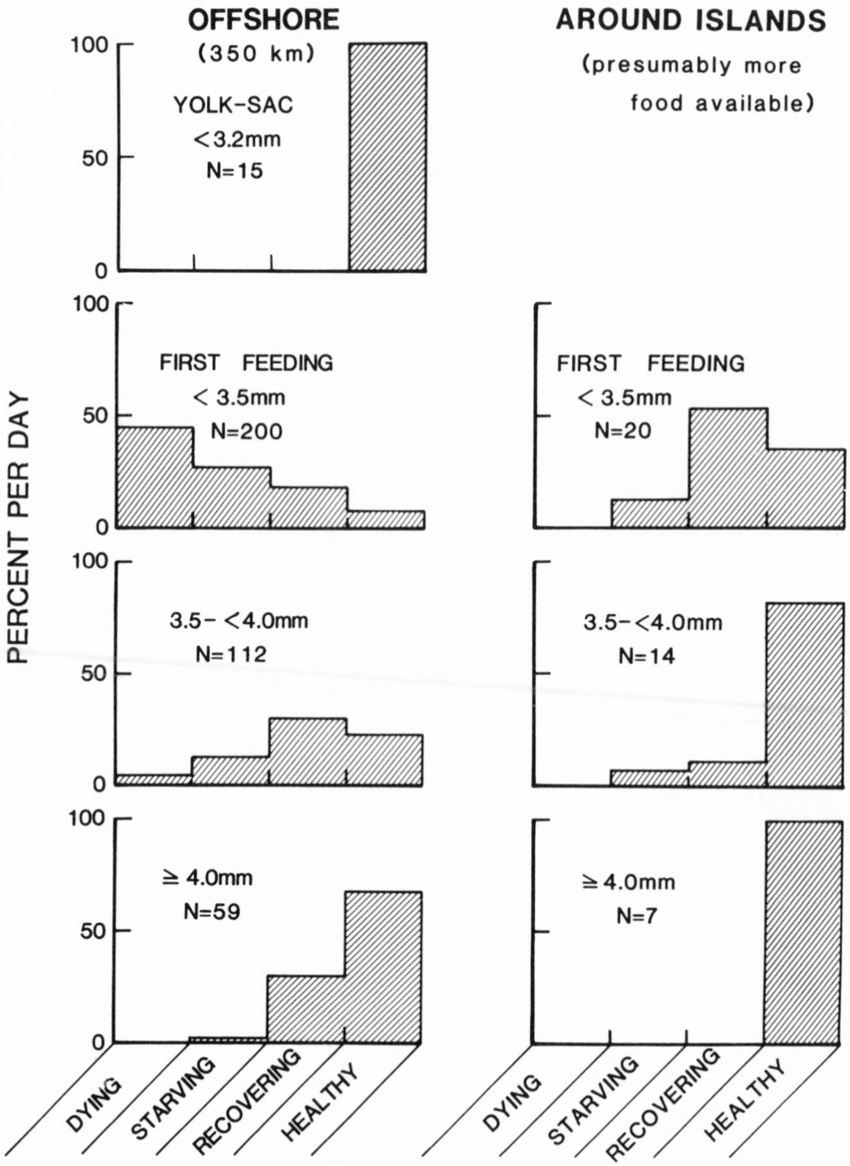

Figure 4.28 *Relative incidence of larval jack mackerel in various nutritional states offshore and around islands in the Southern California Bight. From Theilacker 1986.*

youngest larvae was required because the mesh used to capture them was too large. Starvation was the dominant cause of mortality only for first-feeding larvae (Figure 4.29), and (obviously) only predation could affect the yolksac stage.

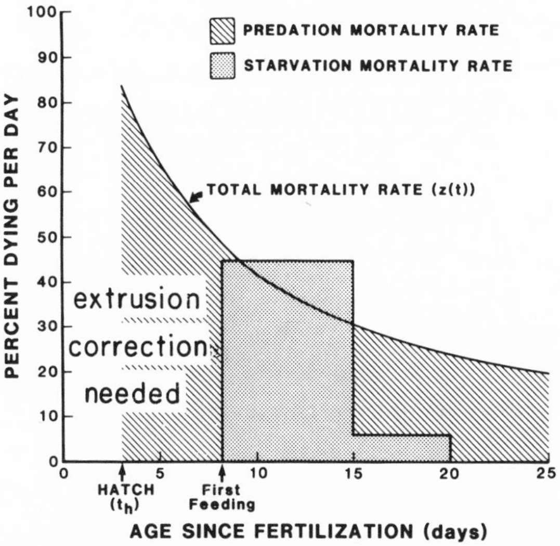

Figure 4.29 *Mortality rate of larval jack mackerel vs. age since fertilization. Total mortality = predation + starvation. Data for youngest larvae are extensively corrected for extrusion through net mesh. From Hewitt et al. 1985.*

Somewhat similar conclusions were reached by Owen et al. (1989) concerning larval anchovy. Although the abundance of small zooplankters (larval food) differed by a factor of 2 at the two sites studied, growth rates of larvae were similar, as were histological indices of larval condition, and the incidence of starvation was very low. The mortality rates were also indistinguishable. Based on abundance of anchovy eggs (from which the adult stock could be computed), Owen et al. argued that cannibalism should have been greater at one site and noncannibalistic predation greater at the other. However, it is unsatisfying (to me, at least) to have to explain the results by invoking a factor that was not directly measured. Indeed, the authors report that the incidence of predatory copepods was highest at the site where most anchovy eggs were found, so that even some indirect evidence weighs against their conclusion.

Speculations about Climatic Change Based on El Niño

The illustrations I have given, moving from large to small and ending in the mesoscale, do not "explain" variation in recruitment to populations of schooling, pelagic fish in the California Current. The elements of a montage are there, and they cover several scales, but the current state of knowledge is not, metaphorically, a unified, artistic creation. At present, the hypothesis that the larval food web is the dominant source of variability in recruitment has been weakened by the failure to explain most of this variability by plausible, physically caused variation in the web Yet, as I have pointed out several times, the food web itself has seldom been sampled adequately. Easily measured bulk properties have usually been the variables available for analysis. Further, few individual studies cover a wide range of scales. We do not really understand how small-scale distributions change in response to large-scale climatic variation, though there are some stimulating possibilities. The simulation models discussed above illustrate another tool for exploring the issue.

The present understanding of the ecology of larval anchovy, though imperfect, suggests some specific questions about how such large-scale climatic change in the ocean, originating in the physical structure, might alter population dynamics of this and ecologically similar species. Qualitatively, it appears that large, nontoxic dinoflagellates—the kind thought by Lasker to be essential for first-feeding anchovy—have a competitive edge over faster-growing diatoms because of their ability to reduce and assimilate nitrate at low intensities of light and, under some conditions, to migrate dielly from the sunlit surface to the nutricline (see Chapter 2). Further, the horizontal distribution and aggregation of dinoflagellates near shore can be affected by the interaction between their diel vertical migration and internal waves that propagate along the thermocline (Kamykowski 1979; see Figure 2.12).

Suppose that, as effects of climatic warming, nutrient-rich subsurface waters form at a reduced rate in polar regions and the nutricline and thermocline deepen in eastern boundary currents. Will the "life-style" and distribution of large dinoflagellates be adversely affected and, if so, will the survival of first-feeding larval anchovy be affected, perhaps benefiting mackerel larvae (which feed immediately on zooplankton)? Or will the larval anchovy happily switch to first-feeding reliance on protozoans, taking advantage of an enhanced microbial loop originating in microflagellates thriving under the new, more oligotrophic physical regime (Figure 4.2)? Or, will there still be dinoflagellates, but only unnutritious or toxic ones?

Even if a Californian El Niño provides a model for warming and greater stratification in the California Current, the 1982–83 El Niño did not provide much support for these speculations concerning anchovy, and too little relevant information was gathered in ways comparable to non-El Niño sampling. Biomass of total phytoplankton in the Southern California Bight was reduced and redistributed relative to the long-term mean pattern, and the thermocline was deeper (McGowan 1985). There is, unfortunately, little information on specific food for larval anchovy, such as the prevalence of offshore, subsurface layers of dinoflagellates. It is known, though, that *Gymnodinium* was present nearshore in spring of 1983, and the assemblage of net phytoplankton there was not anomalous except in October–November, when the water was also anomalously warm (Reid et al. 1985). Preliminary examination of the sediments of the Santa Barbara Basin (Lange et al. 1987) indicates that changes in the diatom flora and radiolarian and foraminiferan faunas can be detected.

However, little information is available on metazoan microzooplankton except through the correlations between micro- and macrozooplankton reported by Smith and Eppley (1982) and between reduced macrozooplanktonic biomass and El Niño events (Wickett 1967, Chelton et al. 1982, McGowan 1985). The spatial scale of longshore patchiness of nearshore zooplankton and net phytoplankton and protozoans was investigated during February–March of the El Niño (Mullin et al. 1989, Reid and Stewart 1989), but few anomalies could be attributed unambiguously to the El Niño, partly because few studies had been conducted in non-El Niño winters using similar methods of sampling and analysis.

The spawning anchovy were individually small, and fecundity per batch of eggs (but not total egg production) was therefore reduced. Larval mortality was elevated, as might be predicted, but this mortality was in the prefeeding, yolksac stage, not the feeding stage (Fiedler et al. 1986). There is little evidence, then, of an effect on the food web of first-feeding larvae in this case. Further, Butler (1989) found that growth rates of surviving larvae were not depressed in 1983 relative to other years, and the 1983 and 1984 year-classes were abundant after recruitment to the fishery. However, juvenile growth rates were depressed and the juveniles were small, so that the 1982–83 El Niño might have had more of an effect on the reproductive potential of those year-classes than on their abundances.

The reduction in macrozooplanktonic biomass that accompanies both El Niño and less dramatic reductions in the southward flow of the California Current and is taken as a measure of the supply of food for larvae, apparently makes little difference to recruitment of Pacific mackerel. Its recruitment success (relative to its spawning biomass) is related directly to the increase in sea level, which signals reduction in southward flow (Figure 4.30). Either the food that is actually critical for mackerel larvae is inversely correlated with bulk planktonic biomass and southward flow, or food is relatively unimportant for survival. Given the general, positive correlation between patchiness and abundance (e.g., Figure 3.21), it is difficult to argue that during El Niño periods adequate food is provided by patchiness rather than by mean conditions.

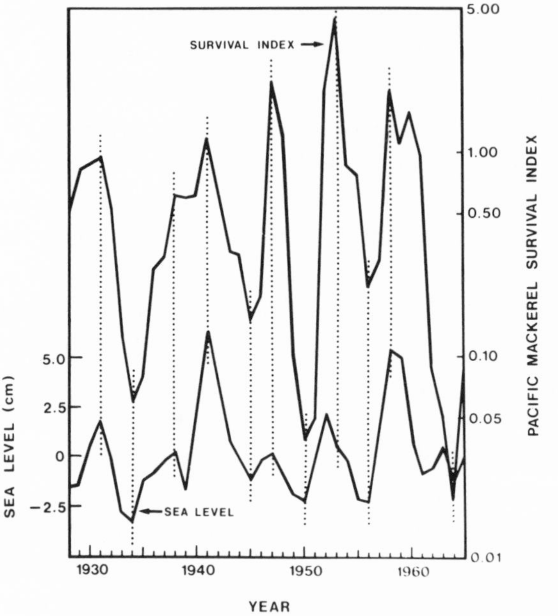

Figure 4.30 *Survival index (age 1 year-class/parental spawning biomass, logarithmic) of Pacific mackerel, and January sea level anomaly (13-month running mean, San Francisco, Los Angeles, and San Diego), 1928–65. From Sinclair et al. 1985. (Prager and MacCall [1988] revised the estimates of spawning biomass, which had been overestimated for 1929–35 and underestimated for 1946–62. This figure does not reflect the revision.)*

Interestingly, recruitment of Pacific mackerel was very poor in 1983 and 1984 (MacCall et al. 1985), following the onset of the very strong El Niño of 1982–84, during which sea level was much higher than normal (McGowan 1985). This suggests that the relation implied by Figure 4.30 may have changed, and may illustrate the quotation from Rothschild (1986) given in Chapter 1.

In some crucial respects, the Californian El Niño may *not* be a good minimodel of the effects of global atmospheric warming. Bakun (1990) argues that, during atmospheric warming, the land will be heated more markedly than the ocean and the increased thermal differential will intensify the equatorward winds that favor upwelling (though Frank et al. [1990] have predicted opposite trends for winds on Canada's east coast).

What will happen if an increase occurs in the duration and intensity of individual upwelling and relaxation events in eastern boundary currents (i.e., fewer events than at present, each lasting longer)? Will entire year-classes be lost to "offshore drift" because a whole season's spawn is exported, or will the succession of less disturbed phytoplankton in the longer stable periods favor large dinoflagellates and thus promote survival of anchovies at the expense of strictly zooplanktivorous larvae? On this same theme, if there are longer periods of wind forcing, will the resulting greater coherence lengths of longshore currents amplify the longshore transport of coastal larvae to topographic regions of semipermanent cross-shelf squirts and jets, where very high offshore velocities will export them? This final suggestion is more pertinent to the Northern California coast, with its stretches of straight coastline broken by capes such as Mendocino, than to the Southern California domain of the anchovy.

Storms could also affect the predators on larvae. Jellyfish (including ctenophores) are relatively fragile, and seem to be more adversely affected by rough weather than piscivorous fish are. If the frequency or duration of calm weather changes, will the nature of the most significant predators change?

Quantitatively, consider the implications of Figure 4.25 from Peterman and Bradford (1987). Suppose that, as a result of climatic changes, the median frequency of 4-day calm periods (on the weighted scale) decreases by 33% from 24 to 16 but the interannual variance in storminess remains constant. Will the median daily mortality rate actually shift to 0.27, as suggested by the regression? Perhaps more important, will the frequency of weak year-classes become so great as to overwhelm the buffer created by iteroparity (Murphy 1967)? This is, of course, a specific example of the general issue discussed in Chapter 1: whether correlations relevant to present climate can be extrapolated to predict the ecological effects of climatic alteration.

5
FURTHER THOUGHTS ON VARIABILITY IN RECRUITMENT

I have attempted to illustrate physical/chemical influences on the distributions of biomass and primary and secondary production of plankton at various scales, and to show how larval fish may be affected by them, possibly (but not necessarily) resulting in variations in recruitment. I have chosen examples in which processes on particular scales present both opportunities for increased understanding and potential impediments (or at least complexities) to be recognized in designing programs of sampling. Finally, I have tried to indicate, in a qualitative and intuitive way, how different scales may be causally connected.

The direct participation of larval fish in the planktonic food web, the relative ease with which they and other plankters may be captured, the capability to rear and study the larvae and their food in the laboratory, and the development of techniques to determine ages and rates of growth have contributed to the elaboration and testing of the long-standing hypothesis that larval feeding ecology is crucial. These examples, however, do not prove that recruitment is determined by the supply of food for young larvae, even for the well-studied anchovy in Californian coastal waters; indeed, the analysis of Peterman et al. (1988) suggests that this is not the case. Nor do the studies of Hewitt et al. (1985), Owen et al. (1989), and Butler (1989) lend strong support even to the simpler hypothesis that larval growth or starvation can readily be related to concurrent abundances of food. Even Theilacker's (1986) result, shown in Figure 4.28, is not strong positive proof, because the abundances of larval food were not determined at the two sites.

The nature of proof is one issue, whether one needs proof only for a specific fishery for a specific period or wants to establish a general principle. Formally, one must attempt to falsify the null hypothesis that there is no relation between recruitment success and supply (or actual use) of larval food, but this falsification can be accomplished by finding a single, statistically significant counterexample. Even this is necessary but not sufficient proof. There should also be data from the same case study that are adequate to test comparable, competing null hypotheses concerning effects of predation, etc., but fail to falsify them. In reality, most ecologists (and fishery managers) would probably be satisfied by inductive proof, namely, a series of cases in which variation in food supply was the most reasonable explanation for recruitment variability, both statistically and mechanistically, even if other plausible explanations had not been eliminated.

Study of interannual variability of recruitment of cod in Norwegian waters (summarized by Ellertsen et al. 1990) provides some strong evidence of the importance of food supply, but it can also be used to illustrate how extensive such investigations can become. Ellertsen et al. found that poor recruitment was statistically associated with anomalously cold years (though a few of the warm years also gave rise to poor year-classes), and that the maximal production of nauplii of copepods (mainly *Calanus finmarchicus*) also occurred later in the spring during cold years. Because naupliar *Calanus* co-occur with larval cod in the upper few tens of meters, and are an important food for them, and because the spawning of cod is relatively invariant in time, Ellertsen et al. proposed that the success of recruitment depends on the coincidence of larval cod and naupliar copepods, and that this is more widespread in warm than in cold years—a match-mismatch argument (Cushing 1972).

What other findings might one add to this scenario to strengthen the argument? First and foremost, one should show a strong correlation between larval survival measured on large scale and subsequent recruitment as demersal juveniles (i.e., a result different from that implied by Figure 4.26 for anchovy). This is distinct from showing only that conditions that *should* enable larvae to survive are correlated with subsequent recruitment.

Ellertsen et al. showed that the fullness of the guts of field-caught larvae depends (nonlinearly) on the concentration of naupliar copepods where the larvae are caught. One would like to be able to show also that larval rates of growth are significantly greater, and mortality less, where and when naupliar copepods are more abundant. This should be true on mesoscales in time and space within years, as well as interannually, and perhaps on small scales as well. Linear regression analysis could be misleading, since the underlying relations are probably curvilinear, as they are for gut fullness; i.e., the larval rates are dependent on naupliar abundances within one range and independent within another range of greater abundances.

For example, the argument could be strengthened if one obtained data for a moderately cold year indicating preferential survival of larvae born late in the season, similar to the data shown for anchovy in Figure 4.19 for 1978, plus evidence that copepod nauplii were significantly rarer in February than in April of that year—plus additional evidence that predation on larvae was not anomalously high in February, and that many larvae were emaciated then. Considering the same example, even if the large-scale mean concentrations of nauplii in February and April were not statistically different, an interesting set of results concerning the importance of still smaller scales would show that the spatial variance was significantly greater in April, that there were nonrandom spatial associations of larvae with patches of anomalously abundant (on this scale) nauplii, and that these larvae were well nourished. Ellertsen et al. in fact considered this possibility with respect to vertical distributions within the upper few tens of meters, but they argued that vertical migrations by larvae and their prey would effectively homogenize the abundances within this region of co-occurrence.

Going beyond correlations on various scales, one should try to demonstrate quantitative balances between supplies and demands. For years of good recruitment, one should be able to show that the naupliar abundances were sufficient, given the larval respiration, defecation, growth, and ability to hunt and capture prey at various ages and temperatures (cf. Figure 4.9), for larvae to meet their metabolic demands. Ellertsen et al. reported the critical concentration as ≥ 50 nauplii per liter (though the field studies of gut fullness suggest that concentrations ≥ 10 nauplii per liter might be sufficient).

Further, one should show that these naupliar abundances were sustained by reproduction in copepod populations long enough, and over large enough areas, to account for subsequent recruitment of juvenile fish, and also to satisfy the needs of all the other populations that compete with larval fish for this food (cf. Dagg et al. 1984, concerning copepod nauplii and larval walleye pollock in the Bering Sea). These competing populations are likely to be diverse and themselves interannually variable. Because of this diversity and the difficulty in separating advective changes in biomass from changes due to differences between input and output of organic matter, it is no trivial exercise to calculate the overall balance between production of and demand for food for larval fish. Indices of growth by individuals in those populations that compete with larval fish (say, a certain size range of chaetognaths) should be positively correlated, from mesoscales to large scales, with recruitment success of the fish, though abundances might not be.

Given that mesoscale differences in the food web are common, successful larvae might well come from specific sites in the spawning range, and these sites might differ from year to year. Assuming that the population was genetically heterogeneous between sites, it might be possible to show correspondence between the genotypic composition of the best-nourished larvae and that of surviving recruits, which would establish a relation between larval nutrition and subsequent recruitment. It could also be important to identify the sites responsible for successful recruitment in years when overall recruitment was strong and in years when it was weak. Identifying and managing sites that sustain a population through poor years is obviously of practical importance.

One would still need to assess the role of predation on prerecruit juveniles as well as larvae, which would be difficult if the abundances of predators and larval food covary. If larval food supply is the primary determinant of recruitment, one should be able to show either that predation is negligible, invariant, or uncorrelated with total larval mortality, or that it is directed primarily against malnourished larvae (i.e., that predation is a proximate more than an ultimate cause of death).

Ellertsen et al. did not try to establish a direct connection between the magnitude or timing of production of copepod nauplii and primary production, nor between those processes and the hydrodynamics affecting them. Verification of the importance of primary production ideally would be based on an analogous set of observations relating unambiguously the variability in production of nauplii by copepods (or production of other larval foods, such as tintinnids) to that of primary production—a truly daunting task!

Predation and cannibalism on juveniles may be at least as significant as larval food supply, particularly for demersal species (e.g., Sissenwine 1986). Because of the difficulty of determining a mortality rate with any precision (since juveniles avoid plankton nets and are often too small to be captured by commercial gear, since determining ages is tedious work, and since mortality rate is a property of populations rather than of individuals), detecting significant *variation* in mortality rates due to predation—the central issue—is another daunting task. The task is much greater than assessing the effects of larval food, because growth rates of individual larvae can be determined and a thorough understanding of variance obtained relatively easily, and it is therefore much easier to test for statistically significant differences in growth between sites or periods than in mortality from predation.

Studies of the degree to which predation on larvae and juveniles affects subsequent recruitment have seldom been conducted in the context of physical and chemical processes in the ocean. This is partly because many of the predators are highly mobile and difficult to sample. Relations between the abundances of piscivorous predators and physical/chemical properties are difficult to establish. An interesting success is the work of Frank and Leggett (1981, 1983). They concluded that the direction of the wind determines whether capelin larvae hatched in Newfoundland bays are exposed to food or to predators. In principle, "webs and scales" are just as important for mortality as for larval feeding, but in the context of mortality they are less readily addressed.

Because of time lags, sampling in a very temporally patchy environment must be not only precise but on the correct scale. Even accurate and precise determination of concurrent abundances in the field of food, larvae, and predators can lead to erroneous conclusions concerning the relative importance of starvation and predation. Taggart and Frank (1990) modeled a situation in which mortality due to starvation was more serious than that due to predation, but the abundance of food integrated over the previous three days affected starvation on any day, while predation was an instantaneous function of the abundance of predators. In this model, the regression of larval mortality rates against the concurrent biomasses of food was not significant, while the comparable regression against the concurrent abundances of predators was significant. This would imply (wrongly) that predation was the more important determinant of variability in recruitment. Because of the mobility and schooling of many predators, the converse error might be made if the investigator failed to integrate the abundance of predators over a larger volume than that occupied by a sample of larvae. The model illustrates an important principle, though it is unrealistic in suggesting that mortality rates can be calculated from field samples with any precision over such small scales in space and time.

I have not discussed at length the advective transport or drift of larvae, though it is clearly due to physical processes. The ecological mechanisms (regulation of growth and death rates) by which transport or drift affects pelagic species usually have not been clearly defined (though demersal shelf species obviously face a problem if the bottom is 2 km beneath them when they metamorphose). Nonetheless, advection also has some scales that are more important than others. For example, the event scale of winter storms is apparently significant for the transport of larval menhaden on the U.S. east coast from offshore spawning areas towards estuarine nursery areas (Checkley et al. 1988), and larvae of several other species must make a similar transit. Conversely, offshore transport of coastal larvae in central and northern California is probably on the mesoscale, spatially, since offshore surface flow is concentrated in topographically determined locations and the intensity is related to event-scale winds (Chapter 4).

More on Scales

To return to a point from Chapter 1, real variation in the size of a population or the structure of an ecosystem in time or space (i.e., variation beyond genuine sampling or analytical imprecision) can provide information about the controls on the population or the functioning of the ecosystem. This is especially true if the variance is known as a function of scale. Given the fact that there are years of strong and weak recruitment, it might clarify the issues to understand more fully the spatial scales on which the differences between years are most marked. That is, strong recruitment in a good year could result from equally improved reproduction and/or survival throughout the population's range (implying a large-scale process as the cause), or from improvement only in specific locations on the mesoscale (implying possibly different causal processes on that scale).

The sketches in Figure 5.1 illustrate this difference, emphasizing it in terms of concentration of food for the species and the resulting rate of growth, based on the assumption (by analogy to an individual) that there are (a) a maintenance concentration below which the population starves (or fails to produce recruits) and (b) a higher concentration at which growth is limited by some factor other than food. Between these, the rate of growth (or recruitment) increases nonlinearly with concentration, and the two will be positively correlated. In both "good" years, the geographical extent of successful recruitment expands and there are regions in which recruitment is maximal. The expansion of range would, in reality, depend not only on supply of food but also on dispersion of adults.

Distinguishing between the two kinds of patterns of good years would reveal the spatial scale of abundant food, and perhaps the underlying physical/chemical forcing, that is most significant to larval growth. As noted above, genetic analysis might be helpful here. Any real ecosystem would, of course, evolve through time from good to bad conditions, with various time lags and other complexities obscuring the relation between the supply of food and growth. In principle, however, the degree of reality of this sketch could be tested by determining the spatial distribution of larval growth rate (e.g., Buckley and Lough 1987; Figures 5.2 and 5.3), its variance within and between sites (which, as shown in Figure 5.4, is a limited form of the scale-specific variance spectrum), and how these differ between years of differing recruitment. Particularly useful are techniques that reveal how growth rate has varied over the larval life span, since if the conditions for rapid growth are on small scales and transient, the larvae growing most rapidly at the time of

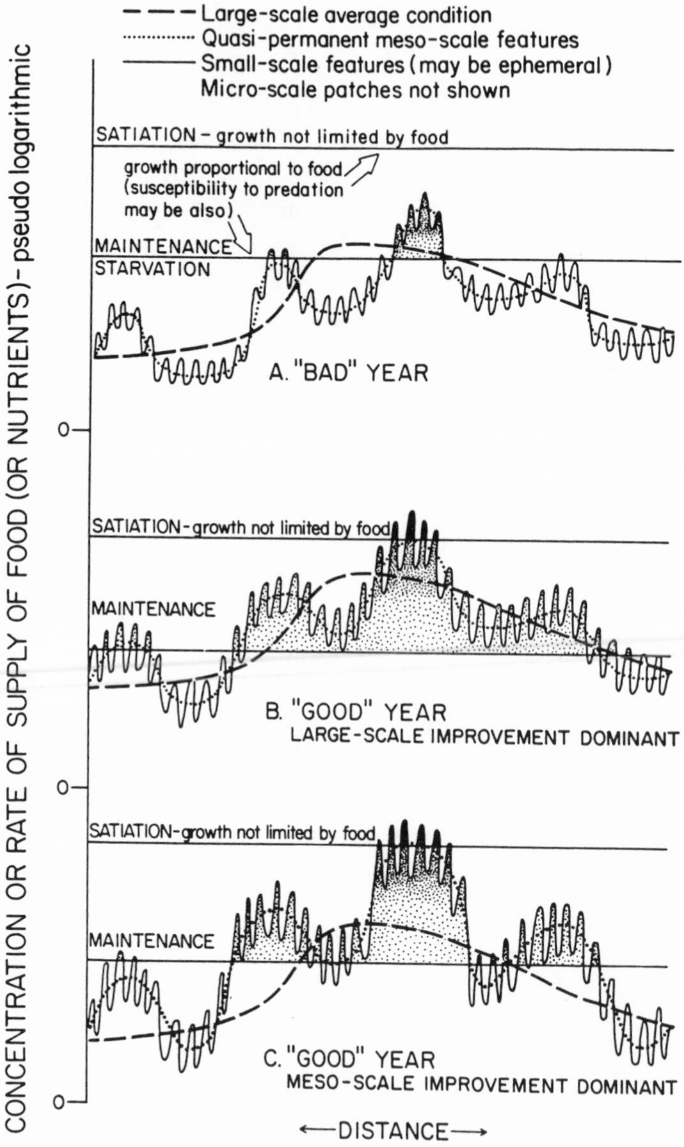

Figure 5.1 *Conceptual sketch of transects across the potential range of a population in one year of poor food supply and two years of abundant food (relative to the amount of food necessary to sustain the population). In year B, the whole area is enriched on the largest scale; in year C, the enrichment is centered in mesoscale areas which were already favorable for growth. Stippling shows areas in which growth is possible, either food-limited (between maintenance and satiation) or food-independent in "good" years. The percentage of the potential range in which growth is possible ranges from 11% for year A to 61% for year B; the percentage in which growth is food-independent ranges from 0% for year A to 11% for year C. From Mullin 1986.*

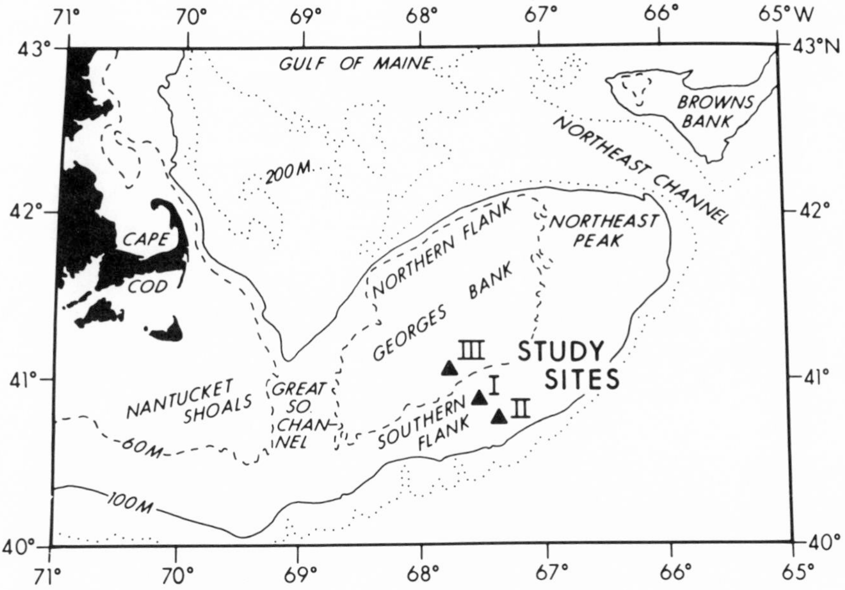

Figure 5.2 Georges Bank, showing depth contours and locations of three study sites. From Buckley and Lough 1987.

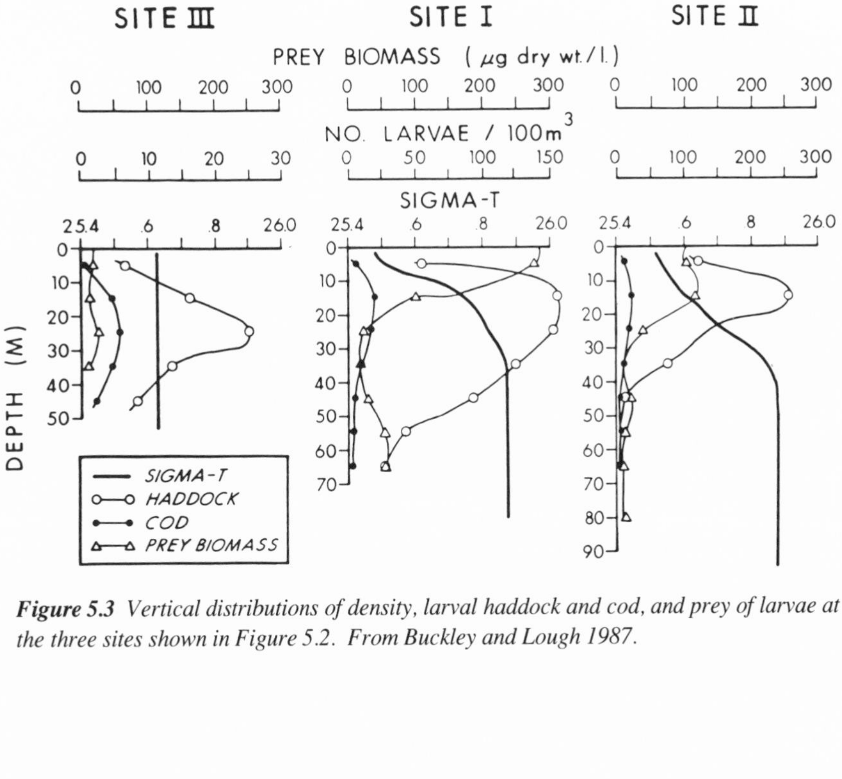

Figure 5.3 Vertical distributions of density, larval haddock and cod, and prey of larvae at the three sites shown in Figure 5.2. From Buckley and Lough 1987.

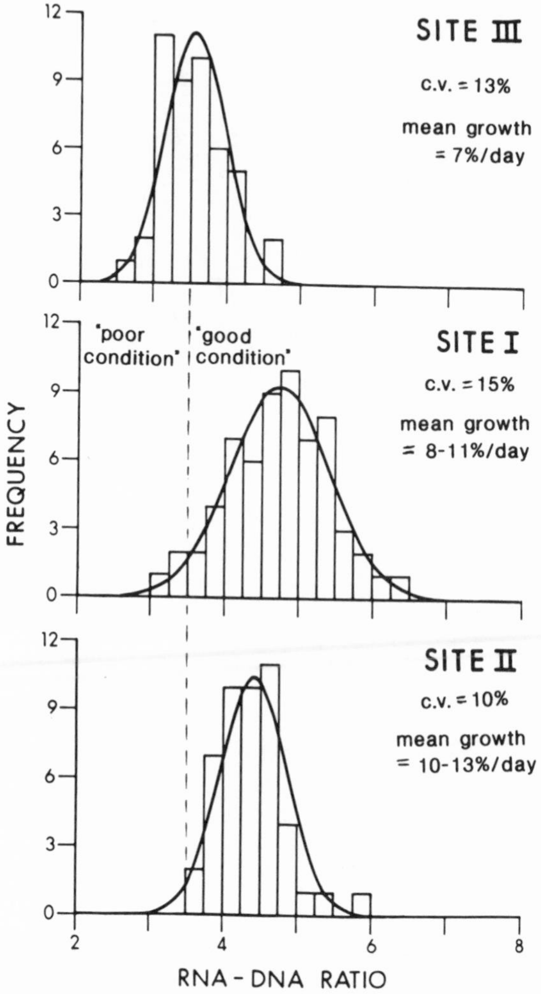

Figure 5.4 Frequency distributions of growth of larval haddock at three sites (see Figure 5.2), as estimated by RNA/DNA ratios, compared with standards of condition determined from laboratory rearing. From Buckley and Lough 1987.

capture may not have had this advantage since first feeding. The widths of otolith daily rings can provide such information, but only for those individuals surviving to be caught by the investigator. If it were possible to recover readable otoliths from the guts of predators on larvae or from sediment traps, an interesting comparison would be possible.

As shown in examples from the California Current System, larval growth is not equivalent to recruitment, which depends also on the spatial distributions of parental reproduction and overall predation. However, if slowly growing larvae are more subject to predation than their more robust siblings (as modeled in Figure 4.20), evenly dispersed predators could enhance numerically the geographical pattern of abundances similar to the pattern shown by the rate of growth of individuals.

Variability, Understanding, and Prediction

In Chapter 1, I expressed some of my opinions concerning different approaches to marine ecology and fisheries oceanography. I now return to fundamental reasons for investigating the scales of variability in space and time of ecological properties and processes affecting food webs and larval fish.

Qualitatively, it is part of the "scientific ideal" to estimate the statistical uncertainty in the mean value of a rate or property so that one can test a null hypothesis such as whether that mean value is the same as another mean value. This principle is often ignored in practice, especially for a process whose value is not measured directly but is calculated from two or more measured, variable parameters: for example, the mean rate of utilization, $\overline{U}$, of a resource of mean concentration, $\overline{R}$, by a consumer at mean concentration, $\overline{C}$, where R and C vary over the space/time domain and are combined by some mathematical transfer function—probably a nonlinear one at that.

Estimates of the variances of $\overline{R}$ and $\overline{C}$ could be carried through the calculation of $\overline{U}$ (cf. Hirota 1974, Platt 1975). Most investigators do not bother to do this, because the parameters are seldom distributed normally, the relation between them is nonlinear, and the resulting variance around U can be horrifyingly large. This last reason is akin to the concatenation of uncertainties I mentioned in Chapter 1. Whether it is an issue of intellectual honesty or common sense is debatable.

Even if this issue can be set aside, the investigator should be concerned about the possibility of erroneously estimating $\overline{U}$ itself. Suppose, for example, that the rate of utilization of R per unit C (= U/C) follows hyperbolic or Michaelis-Menten kinetics (e.g., $U/C = [(U_m R)(K_m + R)^{-1}]$, where U_m is the maximal rate and K_m is the so-called half-saturation constant). If $R << K_m$, the relation becomes essentially linear, and the total rate of uptake, U, is approximately equal:

$$[U_m (K_m)^{-1}] (C R) .$$

Now suppose that R and C covary such that they are positively correlated in time and/or space. (This is reasonable for a consumer and its resource on small scales on the order of the daily foraging volume of individuals, and on very large scales, since the biogeography of a consumer is constrained by that of its resource.) The mean of their products calculated for each point sampled will be greater than the product of their means:

$$U_1 \approx [U_m (K_m)^{-1}](\overline{C_i R_i}) > \bar{U}_2 \approx [U_m (K_m)^{-1}](\bar{C})(\bar{R}) .$$

Because of the covariation and the product, $C_i R_i$, the locations where C_i and R_i are great will contribute importantly to $\bar{U}_1$, which more nearly represents the overall utilization than $\bar{U}_2$ does (cf. Platt and Harrison 1985). The reverse is true if R and C are negatively correlated (which is likely at other, intermediate scales, due to imbalances between the consumer population's rate of utilization and the regeneration rate of the resource). Therefore, even if the investigator cares only about the mean value, $\bar{U}$, there are probably circumstances in which the pattern of covariation of C and R is important. This is akin to the problems of aliasing and biasing I mentioned in Chapter 1.

The stimulus for much of this review is the assumption, which also justifies simple correlation analyses, that particular causes of variation—causes related to the food web—can be identified because they have a dominant and quasi-linear influence, even if only at certain scales, thus permitting inferences concerning relationships. I believe this to be true in many cases, or I would not have written this book, although I also illustrated earlier in this chapter how difficult it can be to obtain unambiguous evidence. However, "understanding," in the sense of knowing causal mechanisms, may be different from successful prediction, and society often is quite content with such prediction, whatever its basis. This perspective suggests that in cases where a sufficient time series of data exists, it can be more efficient to ignore causation and use the past history of variation itself as a basis for prediction.

Various kinds of autocorrelation analyses have been used to search for regularities such as periodicities, since a periodic process is predictable, but most sets of real data are so "noisy" that predictability is possible only in a probabilistic sense. That is, one typically would treat the small-scale variability as random variation ("noise") around some trend on a longer scale, and use this trend to derive a prediction whose precision would depend on the magnitude of the noise. A variance spectrum (see Chapters 1 and 3) is a more elaborate version of this approach, namely, the derivation of a scale-specific description of variation based on the summation of periodicities of different frequencies and amplitudes. One might or might not then try to identify, by correlation, a probable cause for the trend, or different causes for different scales.

A more recent view is that what appears as noise may be a function of a chaotic process at work (chaos, in this restrictive sense, is not randomness, but arises from deterministic, nonlinear relations). For example, Sugihara et al. (1990) have shown that a well-recorded parasite-host system (the human disease, measles) at some scales can be more predictable in the near term when analyzed as a low-dimensional, multiplicative chaos than as (or, in some cases, in addition to) a noisy periodicity. A time series of diatom abundances at Scripps Pier, San Diego, has been treated similarly (Sugihara and May 1990). An important advantage of this approach may be that the problem of nonstationarity (i.e., a change, during the time series, in the cause underlying the time series itself) is less serious in an analysis of chaotic behavior than in autocorrelation (Sugihara et al. 1990).

It is not clear whether data sets of sufficient length and character concerning recruitment will exist to employ this technique in fisheries oceanography. A secondary goal of many traditional correlations is not met: to predict a property that is difficult or expensive to measure from another which is easy and inexpensive. Finally, the advantage of this technique for prediction remains to be demonstrated for data sets in which the variation is dominantly of large scale ("red" rather than "white"), as is the case in the ocean (John H. Steele, Woods Hole Oceanographic Institute, personal communication). Nevertheless, even firm believers in the food web paradigm should consider such alternative approaches to prediction.

LITERATURE CITED

(Chapters in parentheses)

Abbott, M. R., P. J. Richerson, and T. M. Powell. 1982. *In situ* response of phytoplankton fluorescence to rapid variations in light. *Limnology and Oceanography* 27: 218–225. (2)

Alldredge, A. L., and Y. Cohen. 1987. Can microscale chemical patches persist in the sea? Microelectrode study of marine snow, fecal pellets. *Science* 235: 689–691. (2)

Alldredge, A. L., and J. L. Cox. 1982. Primary productivity and chemical composition of marine snow in surface waters of the Southern California Bight. *Journal of Marine Research* 40: 517–527. (2)

Alldredge, A. L., and C. C. Gotschalk. 1990. The relative contribution of marine snow of different origins to biological processes in coastal waters. *Continental Shelf Research* 10: 41–58. (2)

Alldredge, A. L., T. Granata, C. C. Gotschalk, and T. D. Dickey. 1990. The physical strength of marine snow and implications for particle disaggregation in the ocean. *Limnology and Oceanography* 37: 1415–1428. (2)

Alvariño, A. 1980. The relation between the distribution of zooplankton predators and anchovy larvae. *California Cooperative Oceanic Fisheries Investigations Reports* 21: 150–160. (4)

Andrews, J. C. 1983. Deformation of the active space in the low Reynolds number feeding current of calanoid copepods. *Canadian Journal of Fisheries and Aquatic Sciences* 40: 1293–1302. (2)

Arthur, D. K. 1977. Distribution, size, and abundance of microcopepods in the California Current system and their possible influences on the survival of marine teleost larvae. *Fishery Bulletin* (U.S.) 75: 601–612. (4)

Bailey, K. M., and E. D. Houde. 1989. Predation on eggs and larvae of marine fishes and the recruitment problem. *Advances in Marine Biology* 25: 1–83. (4)

Bakun, A. 1986. Definition of environmental variability affecting biological processes in large marine ecosystems. In *Variability and Management of Large Marine Ecosystems,* ed. K. Sherman and L. M. Alexander, AAAS Selected Symposia 99 (Westview Press, Boulder), pp. 89–108. (4)

Bakun, A. 1990. Global climate change and intensification of coastal ocean upwelling. *Science* 247: 198–201. (4)

Barber, R. T., and R. L. Smith. 1981. Coastal upwelling ecosystems. In *Analysis of Marine Ecosystems,* ed. A. R. Longhurst (Academic Press, New York), pp. 31–68. (3)

Barkley, R. A. 1972. Johnston Atoll's wake. *Journal of Marine Research* 30: 201–216. (3)

Barraclough, W. E., R. J. LeBrasseur, and O. D. Kennedy. 1969. Shallow scattering layer in the subarctic Pacific Ocean: Detection by high-frequency echo-sounder. *Science* 166: 611–613. (2)

Beers, J. R. 1986. Organisms and the food web. In *Plankton Dynamics of the Southern California Bight,* ed. R. W. Eppley (Springer-Verlag, Berlin), pp. 84–175. (4)

Beers, J. R., J. D. Trent, F. M. H. Reid, and A. L. Shanks. 1986. Macroaggregates and their phytoplanktonic components in the Southern California Bight. *Journal of Plankton Research* 8: 475–487. (2)

Bishop, J. K. B., M. H. Conte, P. H. Wiebe, M. R. Roman, and C. Langdon. 1986. Particulate matter production and consumption in deep mixed layers: observations in a warm-core ring. *Deep-Sea Research* 33: 1813–1841. (3)

Bjornsen, P. K., and T. G. Nielsen. 1991. Decimeter heterogeneity in the plankton during a pycnocline bloom of *Gymnodinium aureolum. Marine Ecology Progress Series* 73: 263–267. (2, 4)

Boulding, E. G., and T. Platt. 1986. Variation in photosynthetic rates among individual cells of a marine dinoflagellate. *Marine Ecology Progress Series* 29: 199–203. (2)

Boyd, C. M. 1981. Microcosms and experimental planktonic food chains. In *Analysis of Marine Ecosystems,* ed. A. R. Longhurst (Academic, New York), pp. 627–649. (1)

Boyd, C. M. 1985. Is secondary production in the Gulf of Maine limited by the availability of food? *Archiv für Hydrobiologie, Beiheft Ergebnisse der Limnologie* 21: 57–65. (3)

Boyd, S. H., P. H. Wiebe, and J. L. Cox. 1978. Limits of *Nematoscelis megalops* in the Northwestern Atlantic in relation to Gulf Stream cold core rings. II. Physiological and biochemical effects of expatriation. *Journal of Marine Research* 26: 143–159. (3)

Boyd, S. H., P. H. Wiebe, R. H. Backus, J. E. Craddock, and M. A. Daher. 1986. Biomass of the micronekton in Gulf Stream ring 82-B and environs: changes with time. *Deep-Sea Research* 33: 1885–1905. (3)

Buckley, L. J., and R. G. Lough. 1987. Recent growth, biochemical composition, and prey field of larval haddock *(Melanogrammus aeglefinus)* and Atlantic cod *(Gadus morhua)* on Georges Bank. *Canadian Journal of Fisheries and Aquatic Sciences* 44: 14–25. (5)

Bucklin, A., M. M. Reinecker, and C. N. K. Mooers. 1989. Genetic tracers of zooplankton transport in coastal filaments off Northern California. *Journal of Geophysical Research 94*: 8277–8288. (3)

Butler, J. L. 1989. Growth during the larval and juvenile stages of the northern anchovy, *Engraulis mordax,* in the California Current during 1980–84. *Fishery Bulletin* (U.S.) 87: 645–652. (4)

Butler, J. L., and D. Pickett. 1988. Age-specific vulnerability of Pacific sardine, *Sardinops sagax*, larvae to predation by northern anchovy, *Engraulis mordax*. *Fishery Bulletin* (U.S.) 86: 163–167. (4)

Caron, D. A., P. G. Davis, L. P. Madin, and J. McN. Sieburth. 1982. Heterotrophic bacteria and bactivorous protozoa in oceanic microaggregates. *Science* 218: 795–797. (2)

Caron, D. A., P. G. Davis, L. P. Madin, and J. McN. Sieburth. 1986. Enrichment of microbial populations in macroaggregates (marine snow) from surface waters of the North Atlantic. *Journal of Marine Research* 44: 543–565. (2)

Carpenter, S. R. (ed.). 1988. *Complex Interactions in Lake Communities.* Springer-Verlag, Berlin. (1)

Checkley, D. M., Jr., S. Raman, G. L. Maillet, and K. M. Mason. 1988. Winter storm effects on the spawning and larval drift of a pelagic fish. *Nature* 335: 346–348. (5)

Chelton, D. B., P. A. Bernal, and J. A. McGowan. 1982. Large-scale interannual physical and biological interaction in the California Current. *Journal of Marine Research* 40: 1095–1125. (2, 4)

Collie, J. 1988. Book review: Dynamics of Marine Fish Populations. *Natural Resources Modelling* 2: 673–677. (1)

Cowles, T. J., M. R. Roman, A. L. Gauzens, and N. J. Copley. 1987. Short-term changes in the biology of a warm-core ring: Zooplankton biomass and grazing. *Limnology and Oceanography* 32: 653–664. (3)

Cox, J. L., S. Willason, and L. Harding. 1983. Consequences of distributional heterogeneity of *Calanus pacificus* grazing. *Bulletin of Marine Science* 33: 213–226. (2)

Creutzberg, F. 1985. A persistent chlorophyll *a* maximum coinciding with an enriched benthic zone. In *Proceedings Nineteenth European Marine Biology Symposium,* ed. P. E. Gibbs (Cambridge University Press, Cambridge), pp. 97–108. (3)

Currie, D. J. 1984a. Microscale nutrient patches: Do they matter to the phytoplankton? *Limnology and Oceanography* 29: 211–214. (2)

Currie, D. J. 1984b. Phytoplankton growth and the microscale nutrient patch hypothesis. *Journal of Plankton Research* 6: 591–599. (2)

Cushing, D. H. 1972. The production cycle and the numbers of marine fish. In *Conservation and Productivity of Natural Waters,* ed. R. W. Edwards and D. J. Garrod, Symposia of the Zoological Society of London, vol. 29, pp. 213–232. (4, 5)

Cushing, D. H. 1983. Are fish larvae too dilute to affect the density of their food organisms? *Journal of Plankton Research* 5: 847–854. (4)

Cushing, D. H. 1989. A difference in structure between ecosystems in strongly stratified waters and those that are only weakly stratified. *Journal of Plankton Research* 11: 1–13. (4)

Dagg, M. J., and K. D. Wyman. 1983. Natural ingestion rates of the copepods *Neocalanus plumchrus* and *N. cristatus* calculated from gut contents. *Marine Ecology Progress Series* 13: 37–46. (2)

Dagg, M. J., M. E. Clarke, T. Nishiyama, and S. L. Smith. 1984. Production and standing crop of copepod nauplii; food items for the walleye pollock, *Theragra chalcogramma*, in the southeastern Bering Sea. *Marine Ecology Progress Series* 19: 7–16. (5)

Dandonneau, Y., and L. Charpy. 1963. An empirical approach to the island mass effect in the south tropical Pacific based on sea surface chlorophyll concentrations. *Deep-Sea Research* 32: 707–721. (3)

Davis, C. S., and P. H. Wiebe. 1985. Macrozooplankton biomass in a warm-core Gulf Stream ring: time series changes in size structure, taxonomic composition and vertical distribution. *Journal of Geophysical Research* 90: 8871–8884. (3)

Davis, C. S., G. R. Flierl, P. H. Wiebe, and P. J. S. Franks. 1991. Micropatches, turbulence and recruitment in plankton. *Journal of Marine Research* 49: 109–151. (2, 4)

Davoll, P. J., and M. W. Silver. 1986. Marine snow aggregates: life history sequence and microbial community of abandoned larvacean houses from Monterey Bay, California. *Marine Ecology Progress Series* 33: 111–120. (2)

DeAngelis, D. L., and R. M. Cushman. 1990. Potential application of models in forecasting the effects of climate change on fisheries. *Transactions of the American Fisheries Society* 119: 224–239. (1)

Denman, K. L. 1976. Covariability of chlorophyll and temperature in the sea. *Deep-Sea Research* 23: 539–550. (3)

Denman, K. L., and A. E. Gargett. 1983. Time and space scales of vertical mixing and advection of phytoplankton in the upper ocean. *Limnology and Oceanography* 28: 801–815. (2)

Denman, K. L., A. Okubo, and T. Platt. 1977. The chlorophyll fluctuation spectrum in the sea. *Limnology and Oceanography* 22: 1033–1038. (3)

Dickey, T. D. 1988. Recent advances and future directions in multi-disciplinary in situ oceanographic measurement systems. In *Toward a Theory on Biological–Physical Interactions in the World Ocean,* ed. B. J. Rothschild (Kluwer Academic, Dordrecht), pp. 555–598. (1)

Ducklow, H. 1986. Bacterial biomass in warm-core Gulf Stream ring 82-B: mesoscale distributions, temporal changes and production. *Deep-Sea Research* 33: 1789–1812. (3)

Dugdale, R. C., and J. J. Goering. 1967. Uptake of new and regenerated forms of nitrogen in primary productivity. *Limnology and Oceanography* 12: 196–206. (2)

Durbin, A. G., and E. G. Durbin. 1989. Effect of the 'brown tide' on feeding, size and egg laying rate of adult female *Acartia tonsa.* In *Novel Phytoplankton Blooms,* ed. E. M. Cosper, V. M. Bricelj, and E. J. Carpenter (Springer-Verlag, Berlin, pp. 625–646. (3)

Ellertsen, B., P. Fossum, P. Solemdahl, S. Sundby, and S. Taketh. 1990. Environmental influence on recruitment and biomass yields in the Norwegian Sea ecosystem. In *Large Marine Ecosystems: Patterns, Processes, and Yields,* ed. K. Sherman, L. M. Alexander, and B. D. Gold (A.A.A.S., Washington, D.C.), pp. 19–35. (5)

Eppley, R. W. 1981. Autotrophic production of particulate matter. In *Analysis of Marine Ecosystems,* ed. A. R. Longhurst (Academic, New York), pp. 343–361. (2)

Eppley, R. W., and B. J. Peterson. 1979. Particulate organic matter flux and planktonic new production in the deep ocean. *Nature* 282: 677–680. (2)

Eppley, R. W., E. H. Renger, and P. R. Betzer. 1983. The residence time of particulate organic carbon in the surface water of the ocean. *Deep-Sea Research* 30: 311–323. (2)

Evans, R. H., K. S. Baker, O. H. Brown, and R. C. Smith. 1985. Chronology of warm-core ring 82B. *Journal of Geophysical Research* 90: 8803–8811. (3)

Falkowski, P. G. 1984. Physiological responses of phytoplankton to natural light regimes. *Journal of Plankton Research* 6: 295–307. (2)

Falkowski, P. G., and C. D. Wirick. 1981. A simulation model of the effects of vertical mixing on primary productivity. *Marine Biology* 65: 69–75. (2)

Fiedler, P. C. 1982. Zooplankton avoidance and reduced grazing responses to *Gymnodinium splendens* (Dinophyceae). *Limnology and Oceanography* 27: 961–965. (4)

Fiedler, P. C. 1984. Satellite observations of the 1982–83 El Niño along the U.S. Pacific coast. *Science* 224: 1251–1254. (2)

Fiedler, P. C. 1986. Offshore entrainment of anchovy spawning habitat, eggs, and larvae by a displaced eddy in 1985. *California Cooperative Oceanic Fisheries Investigations Reports* 27: 144–152. (4)

Fiedler, P. C., R. D. Methot, and R. P. Hewitt. 1986. Effects of California El Niño 1982–84 on the northern anchovy. *Journal of Marine Research* 44: 317–338. (4)

Flierl, G. R., and J. S. Wroblewski. 1985. The possible influence of warm core Gulf Stream rings upon shelf water larval fish distribution. *Fishery Bulletin* (U.S.) 83: 313–330. (4)

Fogg, G. E. (F.R.S.), B. Egan, G. D. Floodgate, D. A. Jones, J. Y. Kassab, K. Lochter, E. I. S. Rees, S. Scrope-Howe, and C. M. Turley. 1985a. Biological studies in the vicinity of a shallow-sea tidal mixing front. VII. The frontal ecosystems. *Philosophical Transactions of the Royal Society of London* B310: 555–571. (3)

Fogg, G. E. (F.R.S.), B. Egan, S. Hoy, K. Lochte, S. Scrope-Howe, and C. M. Turley. 1985b. Biological studies in the vicinity of a shallow-sea tidal mixing front. I. Physical and chemical background. *Philosophical Transactions of the Royal Society of London* B310: 407–433. (3)

Folkvord, A., and J. R. Hunter. 1986. Size-specific vulnerability of northern anchovy, *Engraulis mordax*, larvae to predation by fishes. *Fishery Bulletin* (U.S.) 84: 859–870. (4)

Fournier, R. O., M. van Det, J. S. Wilson, and N. B. Hargreaves. 1979. Influence of the shelf-break off Nova Scotia on phytoplankton standing stock in winter. *Journal of the Fisheries Research Board of Canada* 36: 1228–1237. (2)

Frank, K. T., and W. G. Leggett. 1981. Wind regulation of emergence times and early larval survival in capelin *(Mallotus vellosus)*. *Canadian Journal of Fisheries and Aquatic Sciences* 38: 215–223. (5)

Frank, K. T., and W. G. Leggett. 1983. Multispecies larval fish associations: accident or adaptation? *Canadian Journal of Fisheries and Aquatic Sciences* 40: 754–762. (5)

Frank, K. T., R. I. Perry, and K. F. Drinkwater. 1990. Predicted response of northwest Atlantic invertebrate and fish stocks to CO_2-induced climate change. *Transactions of the American Fisheries Society* 119: 353–365. (4)

Franks, P. J. S., J. S. Wroblewski, and G. R. Flierl. 1986. Prediction of phytoplankton growth in response to the frictional decay of a warm-core ring. *Journal of Geophysical Research* 91: 7603–7610. (3)

Frey, H. W. (ed.). 1971. *California's Living Marine Resources and Their Utilization.* State of California, Resources Agency, Department of Fish and Game, 148 pp. (4)

Gallegos, C. L., and T. Platt. 1982. Phytoplankton production and water motion in surface mixed layers. *Deep-Sea Research* 29: 65–76. (2)

Gavis, J. 1976. Munk and Riley revisited: nutrient diffusion transport and rates of phytoplankton growth. *Journal of Marine Research* 34: 161–179. (2)

Genin, A., and G. W. Boehlert. 1985. Dynamics of temperature and chlorophyll structures above a seamount: an oceanic experiment. *Journal of Marine Research* 43: 907–924. (3)

Genin, A., L. Haury, and P. Greenblatt. 1988. Interactions of migrating zooplankton with shallow topography: predation by rockfishes and intensification of patchiness. *Deep-Sea Research* 35: 151–175. (3)

Goldman, J. C. 1984. Oceanic nutrient cycles. In *Flows of Energy and Materials in Marine Ecosystems,* ed. M. J. R. Fasham (Plenum. New York), pp. 137–170. (2)

Goldman, J. C., M. R. Dennett, and C. B. Riley. 1981. Marine phytoplankton photosynthesis and transient ammonium availability. *Marine Biology Letters* 2: 323–331. (2)

Gordon, H. R., D. K. Clarke, J. W. Brown, O. B. Brown, and R. H. Evans. 1982. Satellite measurement of the phytoplankton pigment concentration in surface waters of a warm core Gulf Stream Ring. *Journal of Marine Research* 40: 491–502. (3)

Gotschalk, C. C., and A. L. Alldredge. 1989. Enhanced primary production and nutrient regeneration within aggregated diatoms. *Marine Biology* 103: 119–130. (2)

Grenny, W. J., D. A. Bella, and H. Curl. 1973. A theoretical approach to interspecific competition in phytoplankton communities. *American Naturalist* 107: 405–425. (2)

Grice, G. D., and M. R. Reeve (eds.). 1982. *Marine Mesocosms.* Springer-Verlag, Berlin, 430 pp. (1)

Hakanson, J. L. 1987. The feeding condition of *Calanus pacificus* and other zooplankton in relation to phytoplankton pigments in the California Current. *Limnology and Oceanography* 32: 881–894. (3)

Hakanson, J. L. 1989a. Analysis of lipid components for determining the condition of anchovy larvae, *Engraulis mordax. Marine Biology* 102: 143–151. (4)

Hakanson, J. L. 1989b. Condition of larval anchovy *(Engraulis mordax)* in the Southern California Bight, as measured through lipid analysis. *Marine Biology* 102: 153–159. (4)

Hardy, A. C., and E. R. Gunther. 1936. The plankton of the South Georgia whaling grounds and adjacent waters. *Discovery Reports* 11: 1–456. (3)

Harris, G. P. 1984. Phytoplankton productivity and growth measurements: past, present and future. *Journal of Plankton Research* 6: 219–237. (2)

Harris, G. P. 1986. *Phytoplankton Ecology.* Chapman & Hall, New York, 384 pp. (1)

Hassell, M. P., and R. M. May. 1985. From individual behavior to population dynamics. In *Behavioral Ecology: Ecological Consequences of Adaptive Behavior,* ed. R. M. Sibley and R. H. Smith (Blackwell Scientific, Oxford), pp. 3–32. (1)

Haury, L. R. 1984. An offshore eddy in the California Current System. Part IV: Plankton distributions. *Progress in Oceanography* 13: 95–111. (3)

Haury, L. R., M. G. Briscoe, and M. H. Orr, 1979. Tidally generated internal wave packets in Massachusetts Bay. *Nature* 278: 312–317. (2)

Haury, L. R., J. A. McGowan, and P. H. Wiebe. 1978. Patterns and processes in the time–space scales of plankton distributions. In *Spatial Pattern in Plankton Communities,* ed. J. H. Steele (Plenum, New York), pp. 277–327. (1)

Haury, L. R., J. J. Simpson, J. Pelaez, C. J. Koblinsky, and D. Wiesenhahn. 1986. Biological consequences of a recurrent eddy off Point Conception, California. *Journal of Geophysical Research* 91: 12,937–12,956. (3, 4)

Hewitt, R. D., G. H. Theilacker, and N. C. H. Lo. 1985. Causes of mortality in young jack mackerel. *Marine Ecology Progress Series* 26: 1–10. (4)

Heywood, K. J., E. D. Barton, and J. H. Simpson. 1990. The effects of flow disturbance by an oceanic island. *Journal of Marine Research* 48: 55–73. (3)

Hirota, J. 1974. Quantitative natural history of *Pleurobrachia bachei* in La Jolla Bight. *Fishery Bulletin* (U.S.) 72: 295–335. (5)

Hitchcock, G. L., C. Langdon, and T. J. Smayda. 1985. Seasonal variations in the phytoplankton biomass and productivity of a warm-core Gulf Stream ring. *Deep-Sea Research* 32: 1287–1300. (3)

Hitchcock, G. L., C. Langdon, and T. J. Samyda. 1987. Short-term changes in the biology of a Gulf Stream warm-core ring: Phytoplankton biomass and productivity. *Limnology and Oceanography* 32: 919–928. (3)

Holligan, P. M. 1978. Patchiness in subsurface phytoplankton populations on the northwest European continental shelf. In *Spatial Pattern in Plankton Communities,* ed. J. H. Steele (Plenum, New York), pp. 221–238. (3)

Holligan, P. M. 1981. Biological implications of fronts on the northwest European continental shelf. *Philosophical Transactions of the Royal Society of London* A302: 547–562. (3)

Holligan, P. M., R. D. Pingree, and G. T. Mardell. 1985. Oceanic solitons, nutrient pulses, and phytoplankton growth. *Nature* 319: 348–350. (2)

Holligan, P. M., M. Viollier, C. Dupouy, and J. Aiken. 1983. Satellite studies on the distributions of chlorophyll and dinoflagellate blooms in the western English Channel. *Continental Shelf Research* 2: 81–96. (3)

Holligan, P. M., R. P. Harris, R. C. Newell, D. A. Harbour, R. N. Head, E. A. S. Linley, M. I. Lucas, P. R. G. Tranter, and C. M. Weekley. 1984a. Vertical distribution and partitioning of organic carbon in mixed, frontal, and stratified waters of the English Channel. *Marine Ecology Progress Series* 14: 111–127. (3)

Holligan, P. M., P. J. LeB. Williams, D. Purdie, and R. P. Harris. 1984b. Photosynthesis, respiration and nitrogen supply of plankton populations in stratified, frontal, and tidally mixed shelf waters. *Marine Ecology Progress Series* 17: 201–213. (3)

Holloway, G., and K. L. Denman. 1989. Influence of internal waves on primary production. *Journal of Plankton Research* 11: 409–413. (2)

Horne, E. P. W., J. W. Loder, W. G. Harrison, R. Mohn, M. R. Lewis, B. Irwin, and T. Platt. 1989. Nitrate supply and demand at the Georges Bank tidal front. *Sciencia Marina* 53: 145–158. (2, 3)

Horrigan, S. G., and J. J. McCarthy. 1982. Phytoplantkon uptake of ammonium and urea during growth on oxidized forms of nitrogen. *Journal of Plankton Research* 4: 379–389. (2)

Horwood, J. W. 1978. Observations on spatial heterogeneity of surface chlorophyll in one and two dimensions. *Journal of the Marine Biological Association, U.K.* 58: 487–502. (3)

Hunter, J. R., and K. M. Coyne. 1982. The onset of schooling in northern anchovy larvae, *Engraulis mordax. California Cooperative Oceanic Fisheries Investigations Reports* 23: 246–251. (4)

Hunter, J. R., and C. A. Kimbrell. 1980. Early life history of Pacific mackerel *(Scomber japonicus). Fishery Bulletin* (U.S.) 78: 89–101. (4)

Huntley, M., and C. M. Boyd. 1984. Food-limited growth of marine zooplankton. *American Naturalist* 124: 455–478. (3)

Isaacs, J. D., and R. A Schwartzlose. 1965. Migrant sound scatterers: Interaction with the seafloor. *Science* 150: 1810–1813. (3)

Jackson, G. A. 1980. Phytoplankton growth and zooplankton grazing in oligotrophic oceans. *Nature,* 284: 439–441. (2)

Jackson, G. A. 1987. Simulating chemosensory responses of marine microorganisms. *Limnology and Oceanography* 32: 1253–1266. (2)

Jackson, G. A. 1990. A model of the formation of marine algal flocs by physical coagulation processes. *Deep-Sea Research* 37: 1197–1211. (2)

Jamart, B. M., D. F. Winter, K. Banse, G. C. Anderson, and R. K. Lam. 1977. A theoretical study of phytoplankton growth and nutrient distribution in the Pacific Ocean off the northwestern U.S. coast. *Deep-Sea Research* 24: 753–774. (2)

Kamykowski, D. 1979. The growth response of a model *Gymnodinium splendens* in stationary and wavy water columns. *Marine Biology* 50: 289–303. (2, 4)

Kamykowski, D. 1981. The simulation of a Southern California red tide using characteristics of a simultaneously-measured internal wave field. *Ecological Modelling* 12: 253–265. (2)

Kiefer, D. A., and R. Lasker. 1975. Two blooms of *Gymnodinium splendens,* an unarmored dinoflagellate. *Fishery Bulletin* (U.S.) 73: 675–678. (4)

Kingsford, M. J., and J. H. Choat. 1986. Influence of surface slicks on the distribution and onshore movements of small fish. *Marine Biology* 91: 161–171. (2)

Kiørboe, T., and K. Johansen. 1986. Studies on a larval herring *(Clupea harengus* L.) patch in the Buchan area. IV. Zooplankton distribution and productivity in relation to hydrographic features. *Dana* 6: 37–51. (3)

Kiørboe, T., P. Munk, K. Richardson, V. Christensen, and H. Paulsen. 1988. Plankton dynamics and larval herring growth, drift, and survival in a frontal area. *Marine Ecology Progress Series* 44: 205–219. (3)

Knoechel, R., and J. Kalff. 1976a. The applicability of grain density autoradiography to the quantitative determination of algal species production: A critique. *Limnology and Oceanography* 21: 583–590. (2)

Knoechel, R., and J. Kalff. 1976b. Track autoradiography: A method for the determination of phytoplankton species productivity. *Limnology and Oceanography* 21: 590–596. (2)

Lande, R., and C. S. Yentsch. 1988. Internal waves, primary production and the compensation depth of marine phytoplankton. *Journal of Plankton Research* 10: 565–571. (2)

Landry, M. R. 1977. A review of important concepts in the trophic organization of pelagic ecosystems. *Helgoländer wissenschaftliche Meeresuntersuchungen* 30: 8–17. (4)

Lange, C. B., S. K. Burke, and W. H. Berger. 1990. Biological production off southern California is linked to climate change. *Climatic Change* 16: 319–329. (4)

Lange, C. B., W. H. Berger, S. K. Burke, A. E. Casey, A. Schimmelmann, A. Soutar, and A. L. Weinheimer. 1987. El Niño in Santa Barbara Basin: Diatom, radiolarian and foraminiferan responses to the '1983 El Niño' event. *Marine Geology* 78: 153–160. (4)

Lasker, R. 1970. Utilization of zooplankton energy by a Pacific sardine population in the California Current. In *Marine Food Chains,* ed. J. H. Steele (Oliver and Boyd, Edinburgh), pp. 265–284. (4)

Lasker, R. 1975. Field criteria for survival of anchovy larvae: The relation between inshore chlorophyll maximum layers and successful first feeding. *Fishery Bulletin* (U.S.) 73: 453–462. (4)

Lasker, R. 1978. The relation between oceanographic conditions and larval anchovy food in the California Current: Identification of factors contributing to recruitment failure. *Rapports et Procès-verbaux des Réunions, Conseil international pour l'Exploration de la Mer* 173: 212–230. (4)

Lasker, R. 1981. Factors contributing to variable recruitment of the northern anchovy *(Engraulis mordax)* in the California Current: Contrasting years, 1975 through 1978. *Rapports et Procès-verbaux des Réunions, Conseil international pour l'Exploration de la Mer* 178: 375–388. (4)

Lasker, R., and J. R. Zweifel. 1978. Growth and survival of first-feeding northern anchovy larvae *(Engraulis mordax)* in patches containing different proportions of large and small prey. In *Spatial Pattern in Plankton Communities,* ed. J. H. Steele (Plenum, New York), pp. 329–354. (4)

Le Fevre, J. 1986. Aspects of the biology of frontal systems. *Advances in Marine Biology* 23: 163–299. (2, 3)

Lehman, J. T., and D. Scavia. 1982. Microscale patchiness of nutrients in plankton communities. *Science* 216: 729–730. (2)

Lewis, M. R., E. P. W. Horne, J. J. Cullen, N. S. Oakey, and T. Platt. 1984. Turbulent motions may control phytoplankton photosynthesis in the upper ocean. *Nature* 311: 49–50. (2)

Lillelund, K., and R. Lasker. 1971. Laboratory studies of predation by marine copepods on fish larvae. *Fishery Bulletin* (U.S.) 69: 655–667. (4)

Lobel, P. S., and A. R. Robinson. 1986. Transport and entrapment of fish larvae by ocean mesoscale eddies and currents in Hawaiian waters. *Deep-Sea Research* 33: 483–500. (3)

Loder, J. W., and T. Platt. 1985. Physical controls on phytoplankton production at tidal fronts. In *Proceedings Nineteenth European Marine Biology Symposium,* ed. P. E. Gibbs (Cambridge University Press, Cambridge), pp. 3–21. (3)

Loder, J. W., C. K. Ross, and P. C. Smith. 1988. A space- and time-scale characterization of circulation and mixing over submarine banks, with application to the northwestern Atlantic continental shelf. *Canadian Journal of Fisheries and Aquatic Sciences* 45: 1860–1885. (3)

Longhurst, A. R. 1984. Heterogeneity in the ocean: implications for fisheries. *Rapports et Procès-verbaux des Réunions, Conseil international pour l'Exploration de la Mer* 185: 268–282. (1)

MacCall, A. D. 1986. Changes in the biomass of the California Current ecosystem. In *Variability of Management of Large Marine Ecosystems,* ed. K. Sherman and L. M. Alexander, AAAS Selected Symposia 99 (Westview Press, Boulder), pp. 33–54. (4)

MacCall, A. D., R. A. Klingbeil, and R. D. Methot. 1985. Recent increased abundance and potential productivity of Pacific mackerel *(Scomber japonicus). California Cooperative Oceanic Fisheries Investigations Reports* 26: 119–129. (4)

Mackas, D. L., and C. M. Boyd. 1979. Spectral analysis of zooplankton spatial heterogeneity. *Science* 204: 62–64. (3)

Maguire, B., Jr., and W. E. Neill. 1971. Species and individual productivity in phytoplankton communities. *Ecology* 52: 903–907. (2)

Marra, J. 1980. Vertical mixing and primary production. In *Primary Productivity in the Sea,* ed. P. Falkowski, Brookhaven Symposia on Biology 31 (Plenum, New York), pp. 121–137. (2)

Marra, J., and K. Heineman. 1982. Photosynthesis response by phytoplankton to sunlight variability. *Limnology and Oceanography* 27: 1141–1153. (2)

Marra, J., R. W. Houghton, D. C. Boardman, and P. J. Neale. 1982. Variability in surface chlorophyll *a* at a shelf-break front. *Journal of Marine Research* 40: 575–591. (3)

McCarthy, J. J., and J. C. Goldman. 1979. Nitrogenous nutrition of marine phytoplankton in nutrient-depleted waters. *Science* 203: 670–672. (2)

McGowan, J. A. 1985. El Niño 1983 in the Southern California Bight. In *El Niño North: Niño Effects in the Eastern Subarctic Pacific Ocean,* ed. W. P. Wooster and D. L. Fluharty (Washington Sea Grant, Seattle), pp. 166–184. (2, 4)

McGowan, J. A., and C. B. Miller. 1980. Larval fish and zooplankton community structure. *California Cooperative Oceanic Fisheries Investigations Reports* 21: 29–36. (4)

Menzel, D. W., and J. H. Ryther. 1960. The annual cycle of primary production in the Sargasso Sea off Bermuda. *Deep-Sea Research* 6: 351–363. (3)

Methot, R. D., Jr. 1983. Seasonal variation in survival of larval anchovy, *Engraulis mordax*, estimated from the age distribution of juveniles. *Fishery Bulletin* (U.S.) 81: 741–750. (4)

Miller, J. M. 1974. Nearshore distribution of Hawaiian marine fish larvae: effects of water quality, turbidity, and currents. In *The Early Life History of Fish,* ed. J. H. S. Blaxter (Springer-Verlag, Berlin), pp. 217–231. (3)

Mitchell, J. G., and J. A. Fuhrman. 1989. Centimeter scale vertical heterogeneity in bacteria and chlorophyll a. *Marine Ecology Progress Series* 54: 141–148. (2)

Mitchell, J. G., A. Okubo, and J. A. Fuhrman. 1985. Microzones surrounding phytoplankton form the basis for stratified marine microbial ecosystem. *Nature* 316: 58–59. (2)

Mullin, M. M. 1986. Spatial and temporal scales and patterns. In *Plankton Dynamics of the Southern California Bight,* ed. R. W. Eppley (Springer-Verlag, Berlin), pp. 216–273. (2, 5)

Mullin, M. M. 1988. Production and distribution of nauplii and recruitment variability: putting the pieces together. In *Toward a Theory on Biological–Physical Interactions in the World Ocean,* ed. B. J. Rothschild (Kluwer Academic, Dordrecht), pp. 291–320. (3)

Mullin, M. M. 1991. Production of eggs by the copepod *Calanus pacificus* in the southern California sector of the California Current System. *California Cooperative Oceanic Fisheries Investigations Reports* 32: 65–90. (3)

Mullin, M. M., and E. R. Brooks. 1972. The vertical distribution of juvenile *Calanus* (Copepoda) and phytoplankton within the upper 50 m of water off La Jolla, California. In *Biological Oceanography of the Northern North Pacific,* ed. A. Y. Takenouti (Idemitsu Shoten, Tokyo), pp. 347–354. (2)

Mullin, M. M., and E. R. Brooks. 1976. Some consequences of distributional heterogeneity of phytoplankton and zooplankton. *Limnology and Oceanography* 21: 784–796. (2)

Mullin, M. M., E. R. Brooks, and E. F. Stewart. 1989. Nearshore, surface-dwelling zooplanktonic assemblages off Southern California during the anomalous winters of 1983 and 1984. *Continental Shelf Research* 9: 19–36. (4)

Mullin, M. M., E. R. Brooks, F. M. H. Reid, J. M. Napp, and E. F. Stewart. 1985. Vertical structure of nearshore plankton off Southern California: A storm and a larval fish food web. *Fishery Bulletin* (U.S.) 83: 151–170. (2, 4)

Munk, W. H., and G. A. Riley. 1952. Absorption of nutrients by aquatic plants. *Journal of Marine Research* 11: 215–240. (2)

Murphy, G. I. 1967. Vital statistics of the Pacific sardine *(Sardinops caerulea)* and the population consequences. *Ecology* 48: 731–736. (4)

Myers, R. A., and K. Drinkwater. 1989. The influence of Gulf Stream warm core rings on recruitment of fish in the northwest Atlantic. *Journal of Marine Research* 47: 635–656. (4)

Napp, J. M., E. R. Brooks, F. M. H. Reid, P. Matrai, and M. M. Mullin. 1988a. Vertical distribution of marine particles and grazers. I. Vertical distribution of food quality and quantity. *Marine Ecology Progress Series* 50: 45–58. (2)

Napp, J. M., E. R. Brooks, P. Matrai, and M. M. Mullin. 1988b. Vertical distribution of marine particles and grazers. II. Relation of grazer distribution to food quality and quantity. *Marine Ecology Progress Series* 50: 59–72. (2)

Nelson, D. M., J. J. McCarthy, T. M. Joyce, and H. W. Ducklow. 1989. Enhanced near-surface nutrient availability and new production resulting from frictional decay of a Gulf Stream warm-core ring. *Deep-Sea Research* 36: 705–714. (3)

Nelson, D. M., H. W. Ducklow, G. L. Hitchcock, M. A. Brzezinski, T. J. Cowles, C. Garside, R. W. Gould, Jr., T. M. Joyce, C. Langdon, J. J. McCarthy, and C. S. Yentsch. 1985. Distribution and composition of biogenic particulate matter in a Gulf Stream warm-core ring. *Deep-Sea Research* 32: 1347–1369. (3)

Newell, R. C., and E. A. S. Linley. 1984. Significance of microheterotrophs in the decomposition of phytoplankton: estimates of carbon and nitrogen flow based on the biomass of plankton communities. *Marine Ecology Progress Series* 16: 105–119. (3)

Okubo, A. 1978. Horizontal dispersion and critical scales for phytoplankton patches. In *Spatial Pattern in Planktonic Communities,* ed. J. H. Steele (Plenum, New York), pp. 21–42. (3)

O'Neill, R. V. 1988. Hierarchy theory and global change. In *Scales and Global Change,* ed. T. Rosswell, R. G. Woodmansee, and S. G. Riser (Wiley, New York), pp. 29–45. (1)

Ortner, P. B., P. H. Wiebe, L. Haury, and S. Boyd. 1978. Variability in zooplankotn biomass distribution in the northern Sargasso Sea: the contribution of Gulf Stream cold core rings. *Fishery Bulletin* (U.S.) 76: 323–334. (3)

Ortner, P. B., E. M. Hurlburt, and P. H. Wiebe. 1979. Phytohydrography, Gulf Stream rings, and herbivore habitat contrasts. *Journal of Experimental Marine Biology and Ecology* 39: 101–124. (3)

Owen, R. W. 1981. Fronts and eddies in the sea: Mechanisms, interactions, and biological effects. In *Analysis of Marine Ecosystems,* ed. A. R. Longhurst (Academic Press, New York), pp. 197–233. (3)

Owen, R. W. 1989. Microscale and finescale variations of small plankton in coastal and pelagic environments. *Journal of Marine Research* 47: 197–240. (2, 4)

Owen, R. W., N. C. H. Lo, J. L. Butler, G. H. Theilacker, A. Alvariño, J. R. Hunter, and Y. Watanabe. 1989. Spawning and survival patterns of larval northern anchovy, *Engraulis mordax,* in contrasting environments. *Fishery Bulletin* (U.S.) 87: 673–688. (4)

Paffenhöfer, G.-A., and K. D. Lewis. 1990. Perceptive performance and feeding behavior of calanoid copepods. *Journal of Plankton Research* 12: 933–946. (2)

Paffenhöfer, G.-A., J. R. Strickler, and M. Alcaraz. 1982. Suspension-feeding by herbivorous calanoid copepods: A cinematographic study. *Marine Biology* 67: 193–199. (2)

Parrish, R. H., C. S. Nelson, and A. Bakun. 1981. Transport mechanisms and reproductive success of fishes in the California Current. *Biological Oceanography* 1: 175–203. (4)

Payne, A. I. L., J. A. Gulland, and K. H. Brink (eds.). 1987. The Benguela and comparable ecosystems. *South African Journal of Marine Sciences* 5: 1–957. (4)

Peele, E. R., R. E. Murray, R. B. Hanson, L. R. Pomeroy, and R. E. Hodson. 1985. Distribution of microbial biomass and secondary production in a warm-core Gulf Stream ring. *Deep-Sea Research* 32: 1393–1403. (3)

Pepin, P. 1987. Influence of alternative prey abundance on pelagic fish predation on larval fish: a model. *Canadian Journal of Fisheries and Aquatic Sciences* 44: 222–227. (4)

Peterman, R. M., and M. J. Bradford. 1987. Wind speed and mortality rate of a marine fish, the northern anchovy *(Engraulis mordax). Science* 235: 354–356. (4)

Peterman, R. M., M. J. Bradford, N. C. H. Lo, and R. D. Methot. 1988. Contribution of early life stages to interannual variability in recruitment of northern anchovy *(Engraulis mordax). Canadian Journal of Fisheries and Aquatic Sciences* 45: 8–16. (4, 5)

Peterson, W. T., P. Tiselius, and T. Kiørboe. 1991. Copepod egg production, moulting and growth rates, and secondary production, in the Skagerrak in August 1988. *Journal of Plankton Research* 13: 131–154. (3)

Pieper, R. E., D. V. Holliday, and G. S. Kleppel. 1990. Quantitative zooplankton distributions from multifrequency acoustics. *Journal of Plankton Research* 12: 433–441. (3)

Pingree, R. D. 1978. Mixing and stabilization of phytoplankton distributions on the northwest European continental shelf. In *Spatial Pattern in Plankton Communities,* ed. J. H. Steele (Plenum, New York), pp. 181–220. (3)

Pingree, R. D., P. M. Holligan, and R. N. Head. 1977. Survival of dinoflagellate blooms in the western English Channel. *Nature* 265: 266–269. (3)

Platt, T. 1975. Analysis of the importance of spatial and temporal heterogeneity in the estimation of annual production by phytoplankton in a small, enriched, marine basin. *Journal of Experimental Marine Biology and Ecology* 18: 99–110. (5)

Platt, T. 1978. Spectral analysis of spatial structure in phytoplankton populations. In *Spatial Structure in Plankton Communities,* ed. J.. H. Steele (Plenum, New York), pp. 73–84. (3)

Platt, T., and K. L. Denman. 1975. A general equation for the mesoscale distribution of phytoplankton in the sea. *Mémoires de la Société Royale de Liège* 7: 31–42. (3)

Platt, T., and W. G. Harrison. 1985. Biogenic fluxes of carbon and oxygen in the ocean. *Nature* 318: 55–58. (5)

Power, J. H. 1986. A model of drift of the northern anchovy, *Engraulis mordax,* larvae in the California Current. *Fishery Bulletin* (U.S.) 84: 585–603. (4)

Prager, M. H., and A. D. MacCall. 1988. Revised estimates of historical spawning biomass of the Pacific mackerel, *Scomber japonicus. California Cooperative Oceanic Fisheries Investigations Reports* 29: 81–90. (4)

Prezelin, B. B., and A. L. Alldredge. 1983. Primary production of marine snow during and after an upwelling event. *Limnology and Oceanography* 28: 1156–1167. (2)

Price, H. J., G.-A. Paffenhöfer, and J. R. Strickler. 1983. Modes of cell capture in calanoid copepods. *Limnology and Oceanography* 28: 116–123. (2)

Putt, M., and B. B. Prezelin. 1985. Observations of diel patterns of photosynthesis in cyanobacteria and nanoplankton in the Santa Barbara Channel during "el Niño." *Journal of Plankton Research* 7: 779–790. (2)

Radovich, J. 1982. The collapse of the California sardine fishery: what have we learned? *California Cooperative Oceanic Fisheries Investigations Reports* 23: 56–78. (4)

Reid, F. M. H. and E. Stewart. 1989. Nearshore microplanktonic assemblages off southern California in February 1983 during the El Niño event. *Continental Shelf Research* 9: 37–50. (4)

Reid, F. M. H., C. B. Lange, and M. M. White. 1985. Microplankton species assemblages at Scripps Pier from March to November 1983 during the 1982–1984 El Niño event. *Botanica Marina* 28: 443–452. (4)

Reid, J. L., Jr. 1962. On circulation, phosphate–phosphorus content, and zooplankton volumes in the upper part of the Pacific Ocean. *Limnology and Oceanography* 7: 287–306. (2)

Richardson, K., M. R. Heath, and N. J. Pihl. 1986a. Studies of a larval herring *(Clupea harengus* L.) patch in the Buchan area. I. The distribution of larvae in relation to hydrographic features. *Dana* 6: 1–10. (3)

Richardson, K., M. R. Heath, and S. M. Pedersen. 1986b. Studies of a larval herring *(Clupea harengus* L.) patch in the Buchan area. III. Phytoplankton distribution and primary productivity. *Dana* 6: 25–36. (3)

Richardson, P. L. 1983. Gulf Stream rings. In *Eddies in Marine Science,* ed. A. R. Robinson (Springer-Verlag, Berlin), pp. 19–45. (3)

Richerson, P., R. Armstrong, and C. R. Goldman. 1970. Contemporaneous disequilibrium, a new hypothesis to explain the 'paradox of the plankton.' *Proceedings of the National Academy of Sciences* 67: 1710–1714. (1)

Riley, G. A. 1946. Factors controlling phytoplankton populations of Georges Bank. *Journal of Marine Research* 6: 54-73. (1)

Roesler, C. S., and D. B. Chelton. 1987. Zooplankton variability in the California Current, 1951–1982. *California Cooperative Oceanic Fisheries Investigations Reports* 28: 59–96. (2)

Roman, M. R., A. L. Gauzens, and T. J. Cowles. 1985. Temporal and spatial changes in epipelagic microzooplankton and mesozooplankton biomass in warm-core Gulf Stream ring 82-B. *Deep-Sea Research* 32: 1007–1022. (3)

Roman, M. R., C. S. Yentsch, A. L. Gauzens, and D. A. Phinney. 1986. Grazer control of the fine-scale distribution of phytoplankton in warm-core Gulf Stream rings. *Journal of Marine Research* 44: 795–813. (3)

Rothschild, B. J. 1986. *Dynamics of Marine Fish Populations.* Harvard, Cambridge, 277 pp. (1)

Rothschild, B. J., and T. R. Osborn. 1988. Small-scale turbulence and plankton contact rates. *Journal of Plankton Research* 10: 465–474. (2, 3)

Saitoh, S., S. Kosaka, and J. Iisaka. 1986. Satellite infrared observations of Kuroshio warm-core rings and their application to study of Pacific saury migration. *Deep-Sea Research* 33: 1601–1615. (3)

Savidge, G. 1980. Photosynthesis of marine phytoplankton in fluctuating light regimes. *Marine Biology Letters* 1: 295–300. (2)

Scheiber, H. N. 1990. California marine research and the founding of modern fisheries oceanography: CalCOFI's early years, 1947–1968. *California Cooperative Oceanic Fisheries Investigations Reports* 31: 63–83. (4)

Schoener, T. W. 1986. Mechanistic approaches to community ecology: a new reductionism? *American Zoologist* 26: 81–106. (1)

Shanks, A. L. 1983. Surface slicks associated with tidally forced internal waves may transport pelagic larvae of benthic invertebrates and fishes shoreward. *Marine Ecology Progress Series* 13: 311–315. (2)

Shanks, A. L. 1988. Further support for the hypothesis that internal waves can cause shoreward transport of larval invertebrates and fish. *Fishery Bulletin* (U.S.) 86: 703–714. (2)

Shanks, A. L., and J. D. Trent. 1979. Marine snow: microscale nutrient patches. *Limnology and Oceanography* 24: 850–854. (2)

Shanks, A. L., and W. G. Wright. 1987. Internal-wave-mediated shoreward transport of cyprids, megalopae, and gammarids and correlated longshore differences in settling rate of intertidal barnacles. *Journal of Experimental Marine Biology and Ecology* 114: 1-13. (2)

Shulenberger, E., and J. L. Reid, Jr. 1981. The Pacific shallow oxygen maximum, deep chlorophyll maximum, and primary productivity reconsidered. *Deep-Sea Research* 28: 901–920. (2)

Simpson, J. H. 1981. The shelf–sea fronts: implications of their existence and behaviour. *Philosophical Transactions of the Royal Society of London* A302: 531–546. (3)

Simpson, J. H., P. B. Tett, M. L. Argoti-Espinoza, A. Edwards, K. J. Jones, and G. Savidge. 1982. Mixing and phytoplankton growth around an island in a stratified sea. *Continental Shelf Research* 1: 15–31. (3)

Simpson, J. J. 1987. Transport processes affecting the survival of pelagic fish stocks in the California Current. *American Fisheries Society Symposia* 2: 39–60. (3, 4)

Simpson, J. J., C. L. Koblinsky, J. Pelaez, L. R. Haury, and D. Wiesenhahn. 1986. Temperature–plant pigment–optical relations in a recurrent offshore mesoscale eddy near Point Conception, California. *Journal of Geophysical Research* 91: 12,919–12,936. (3)

Sinclair, M. 1988. *Marine Populations.* Washington Sea Grant, Seattle, 252 pp. (3, 4)

Sinclair, M., M. J. Trombley, and P. Bernal. 1985. El Niño events and variability in a Pacific mackerel *(Scomber japonicus)* survival index: support for Hjort's second hypothesis. *Canadian Journal of Fisheries and Aquatic Sciences* 42: 602–608. (4)

Sissenwine, J. P. 1986. Perturbation of a predator-controlled continental shelf ecosystem. In *Variability and Management of Large Marine Ecosystems,* ed. K. Sherman and L. M. Alexander, AAAS Selected Symposia 99 (Westview Press, Boulder), pp. 55–85. (5)

Smith, P. E. 1978. Biological effects of ocean variability: Time and space scales of biological response. *Rapports et Procès-verbaux des Réunions, Conseil international pour l'Exploration de la Mer* 173: 117–127. (4)

Smith, P. E., and R. W. Eppley. 1982. Primary production and the anchovy population in the Southern California Bight: Comparison of time series. *Limnology and Oceanography* 29: 1–17. (4)

Soutar, A., and J. D. Isaacs. 1974. Abundance of pelagic fish during the 19th and 20th centuries as recorded in anaerobic sediment off the Californias. *Fishery Bulletin* (U.S.) 72: 257–274. (4)

Star, J. L., and J. L. Cullen. 1981. Spectral analysis: A caveat. *Deep-Sea Research* 28: 93–97. (3)

Star, J. L., and M. M. Mullin. 1981. Zooplankton assemblages in three areas of the North Pacific as revealed by continuous horizontal transects. *Deep-Sea Research* 28: 1303–1322. (3)

Steele, J. H. 1978. Some comments on plankton patches. In *Spatial Patterns in Plankton Communities,* ed. J. H. Steele (Plenum, New York), pp. 1–20. (1)

Steele, J. H. 1985. A comparison of terrestrial and marine ecological systems. *Nature* 313: 355–358. (1)

Steele, J. H., and E. W. Henderson. 1981. A simple plankton model. *American Naturalist* 117: 676–691. (1)

Strickler, J. R. 1982. Calanoid copepods, feeding currents, and the role of gravity. *Science* 218: 158–160. (2)

Strickler, J. R. 1984. Sticky water: a selective force in copepod evolution. In *Trophic Interactions within Aquatic Ecosystems,* ed. D .G. Meyers and J. R. Strickler, AAAS Selected Symposia 85 (Westview Press, Boulder), pp. 187–239. (2)

Sugihara, G., and R. M. May. 1990. Nonlinear forecasting as a way of distinguishing chaos from measurement error in time series. *Nature* 344: 734–741. (5)

Sugihara, G., B. Grenfell, and R. M. May. 1990. Distinguishing error from chaos in ecological time series. *Philosophical Transactions of the Royal Society of London* B330: 235–251. (5)

Sundby, S., and P. Fossum. 1990. Feeding conditions of Arcto-norwegian cod larvae compared with the Rothschild–Osborn theory on small-scale turbulence and plankton contact rates. *Journal of Plankton Research* 12: 1153–1162. (2, 4)

Taggart, C. T., and K. T. Frank. 1990. Perspectives on larval fish ecology and recruitment processes: Probing the scales of relationships. In *Large Marine Ecosystems: Patterns, Processes, and Yields,* ed. K. Sherman, L. M. Alexander, and B. D. Gold (A.A.A.S., Washington, D.C.), pp. 151–164. (5)

Theilacker, G. H. 1986. Starvation-induced mortality of young sea-caught jack mackerel, *Trachurus symmetricus,* determined with histological and morphological methods. *Fishery Bulletin* (U.S.) 84: 1–17. (4)

Thomas, W. H., and C. H. Gibson. 1990a. Effects of small-scale turbulence on microalgae. *Journal of Applied Phycology* 2: 71–77. (2)

Thomas, W. H., and C. H. Gibson. 1990b. Quantified small-scale turbulence inhibits a red tide dinoflagellate, *Gonyaulax polyedra* Stein. *Deep-Sea Research* 37: 1583–1593. (2)

Tiselius, P., T. G. Nielsen, G. Brueul, A. Jaanus, A. Korshenko, and Z. Witek. 1991. Copepod egg production in the Skagerrak during SKAGEX, May–June 1990. *Marine Biology* 111: 445–453. (3)

Turpin, D. H., and P. J. Harrison. 1979. Limiting nutrient patchiness and its role in phytoplankton ecology. *Journal of Experimental Marine Biology and Ecology* 39: 151–166. (2)

Uda, M., and M. Ishino. 1958. Enrichment pattern resulting from eddy systems in relation to fishing grounds. *Journal of the Tokyo University of Fisheries* 44: 105–119. (3)

Venrick, E. L. 1988. The vertical distribution of chlorophyll and phytoplankton species in the North Pacific central environment. *Journal of Plankton Research* 10: 987–998. (2)

Vlymen, W. J. 1977. A mathematical model of the relationship between larval anchovy *(Engraulis mordax)* growth, prey microdistribution, and larval behavior. *Environmental Biology of Fishes* 2: 211–233. (4)

Walsh, P., and L. Legendre. 1988. Photosynthetic responses of the diatom *Phaeodactylum tricornutum* to high frequency light fluctuations simulating those induced by sea surface waves. *Journal of Plankton Research* 10: 1077–1082. (2)

Weber, L. H., S. Z. El-Sayed, and I. Hampton. 1986. The variance spectra of phytoplankton, krill, and water temperature in the Antarctic Ocean south of Africa. *Deep-Sea Research* 33: 1329–1343. (3)

Wickett, W. P. 1967. Ekman transport and zooplankton concentration in the North Pacific Ocean. *Journal of the Fisheries Research Board of Canada* 24: 581–594. (4)

Wiebe, P. H., and S. H. Boyd. 1978. Limits of *Nematoscelis megalops* in the Northwestern Atlantic in relation to Gulf Stream cold core rings. I. Horizontal and vertical distribution. *Journal of Marine Research* 36: 119–142. (3)

Wiebe, P. H., E. M. Hurlburt, E. J. Carpenter, A. E. Jahn, G. P. Knapp III, S. H. Boyd, P. B. Ortner, and J. L. Cox. 1976. Gulf Stream cold core rings: Large scale interaction sites for open ocean plankton communities. *Deep-Sea Research* 23: 695–710. (3)

Wiebe, P. H., V. Barber, S. H. Boyd, C. Davis, and G. R. Flierl. 1985. Macrozooplankton biomass in Gulf Stream warm-core rings: Spatial distribution and temporal change. *Journal of Geophysical Research* 90: 8885–8901. (3)

Wolf, K. U., and J. D. Woods. 1988. Lagrangian simulation of primary production in the physical environment: the deep chlorophyll maximum and nutricline. In *Toward a Theory on Biological–Physical Interactions in the World Ocean,* ed. B. J. Rothschild (Kluwer Academic, Dordrecht), pp. 51–70. (2)

Wroblewski, J. S., and J. G. Richman. 1987. The non-linear response of plankton to wind mixing events: implications for the survival of larval northern anchovy. *Journal of Plankton Research* 9: 103–123. (2, 4)

Wroblewski, J. S., J. G. Richman, and G. L. Mellot. 1989. Optimal wind conditions for the survival of larval northern anchovy, *Engraulis mordax:* a modeling investigation. *Fishery Bulletin* (U.S.) 87: 387–398. (4)

Yamamoto, T., and S. Nishizawa. 1986. Small-scale zooplankton aggregations at the front of a Kuroshio warm-core ring. *Deep-Sea Research* 33: 1729–1740. (3)

Yamazaki, H., and T. R. Osborn. 1988. Review of oceanic turbulence: implications for biodynamics. In *Toward a Theory on Biological–Physical Interactions in the World Ocean,* ed. B. J. Rothschild (Kluwer Academic, Dordrecht), pp. 215–234. (2)

Yamazaki, H., T. R. Osborn, and K. D. Squires. 1991. Direct numerical simulation of planktonic contact in turbulent flow. *Journal of Plankton Research* 13: 629–643. (2)

Zeldis, J. R., and J. B. Jillett. 1982. Aggregation of pelagic *Munida gregaria* (Fabricus) (Decapoda, Anomura) by coastal fronts and internal waves. *Journal of Plankton Research* 4: 839–857. (2)

INDEX